The Understanding, Prevention and Control of Human Cancer

History of Science and Medicine Library

VOLUME 49

The titles published in this series are listed at *brill.com/hsml*

The Understanding, Prevention and Control of Human Cancer

The Historic Work and Lives of Elizabeth Cavert Miller and James A. Miller

By

Robert G. McKinnell

BRILL

LEIDEN | BOSTON

Cover illustration: Image #SO 5536 courtesy of the University of Wisconsin Archives.

Library of Congress Cataloging-in-Publication Data

McKinnell, Robert Gilmore.
The understanding, prevention and control of human cancer : the historic work and lives of Elizabeth Cavert Miller and James A. Miller / by Robert G. McKinnell.
pages cm. — (History of science and medicine library, ISSN 1872-0684 ; volume 49)
Includes bibliographical references and index.
ISBN 978-90-04-28679-5 (hardback : acid-free paper) — ISBN 978-90-04-28680-1 (e-book) 1. Miller, Elizabeth C. 2. Miller, James A. (James Alexander), 1915–2000. 3. Oncologists—United States—Biography. 4. Cancer—Research—United States—History—20th century. 5. Cancer—Prevention—Research—United States—History—20th century. 6. Chemical carcinogenesis—Research—United States—History—20th century. 7. Environmentally induced diseases—Research—United States—History—20th century. 8. Cancer—Environmental aspects—Research—United States—History—20th century. I. Title.

RC265.8.M55M35 2016
362.19699'40092—dc23
[B]

2015034777

This publication has been typeset in the multilingual "Brill" typeface. With over 5,100 characters covering Latin, IPA, Greek, and Cyrillic, this typeface is especially suitable for use in the humanities.
For more information, please see www.brill.com/brill-typeface.

ISSN 1872-0684
ISBN 978-90-04-28679-5 (hardback)
ISBN 978-90-04-28680-1 (e-book)

This book is printed on acid-free paper.

For

Beverly Kerr McKinnell
who encouraged me in my desire to begin this book
and for her years of fervent support

and

Suzanne Madison Rohman
who gently urged me to finish *this work*
and who contributed to that endeavor in so many ways

Contents

Acknowledgements

First and foremost, I must thank the Cavert and Miller families who told me about Elizabeth Cavert Miller and James Alexander Miller and who made written family histories, correspondence and photographs freely available.

While teaching my course on the Biology of Cancer at the University of Minnesota, I discovered that Elizabeth Cavert grew up in my neighborhood of Saint Anthony Park in Saint Paul, Minnesota. Many years ago, I met her father in a neighborhood association. Later I found that her brother, Dr. Henry Mead Cavert, also a Saint Anthony Park neighbor, a friend, and a Professor of Physiology Emeritus and Associate Dean Emeritus of the University of Minnesota Medical School, would be willing to discuss his family and memories of his sister Elizabeth. Mead told me of Elizabeth and James' daughters, Linda Miller Forbess and Helen Miller Alexander. Linda, of Schofield, Wisconsin, who might be described as the archivist of the family, was generous of her time and resources. I had the pleasure of meeting with Helen, a professor at the University of Kansas, Lawrence, Kansas several times. I met Elizabeth's sister Jane and her husband, John E. Mitchell, in their apartment in Madison, Wisconsin, just months before their deaths. The Mitchells provided information about the young Elizabeth that was available from no other source.

I had the good fortune to meet with Roberta Miller Williams of Pittsburgh, Pennsylvania, a niece of James Miller, who provided me with Miller history and graciously guided me to the childhood home of James. I have no words that will adequately express my appreciation for the sustained help of these family members.

Charlton, New York, is the home of many of Elizabeth's antecedents. I traveled to Charlton to obtain a feeling for the town and surrounding area that was so important to the history of her family. I was helped in this endeavor by several citizens of Charlton including the town historian Laura Linder. Mari and John Fettke provided introductions to people who assisted me. Connie Wood of the Eildon Tweed Farm was generous with information about Carlton.

Colleagues, John M. Eyler, Ph.D. and Jennifer L. Gunn, Ph.D., Program in the History of Medicine, University of Minnesota Medical School, provided advice and guidance. I am especially indebted to Jole Shackelford, Ph.D., also of the History of Medicine Program, who provided encouragement for several years. Fortuitously, Jole arranged for me to meet Michiel S. Thijssen, of Brill. Thank you Jole!

Early versions of the manuscript were read by Helen Miller Alexander, Ph.D., H. Mead Cavert, M.D., Ph.D., Marie A. DiBerardino, Ph.D., Nancy M. Finney,

Linda Miller Forbess, Kristine Klos, Ph.D., Alan O. Perantoni, Ph.D., Suzanne Madison Rohman and John F. Van Pilsum, Ph.D. My long time friend C. C. Clawson, M.D., read and re-read the manuscript and with his creative suggestions significantly enhanced this book. A special note of appreciation goes to the late Allan H. Conney, Ph.D. who studied and collaborated with the Millers. Dr. Conney read the manuscript, made useful suggestions to enhance this book and encouraged me in my endeavor.

I acknowledge with great appreciation the multiple and careful readings with suggested changes in prose by my long time good friend and colleague Debra Louise Carlson, Ph.D. Her irreplaceable role in this project can be described with no other term than priceless. Deb, I will always be in your debt.

Curator Emeritus Elaine M. Challacombe, M.L.S. and current Curator Lois G. Hendrickson, M.L.S. of the Owen H. Wangensteen Historical Library of Biology and Medicine, University of Minnesota, were gracious with words of encouragement and support. Christopher Herzberg, also of the Wangensteen, was *always* there when I needed his help. I am afraid that I took too much of your time, Chris, but that time and your help will always be much appreciated.

I thank James V. Curley, National Museum of Health and Medicine, Washington, D.C., Emily Reimer, Randy Brooks, Anders Asp, Chacko T. Kuruvilla and Steven E. Irons of the BioMedical Library, University of Minnesota, and Susan Kerr McKinnell, Senior Technology Training Consultant, Office of Information Technology, University of Minnesota, for their continued attention and prompt remedial actions regarding the inadequacies of my computer and word-processing skills.

Executive Director Todd Mahon, formerly of the Anoka County Historical Society, provided me with access to archived copies of the *Anokan*, the yearbook of Anoka High School, with its records of Elizabeth Cavert's high school accomplishments.

I thank Vicki L. Tobias and Catherine Jacob for their help in finding useful illustrations (the cover of this book is an example) and biographical data in the University Archives, University of Wisconsin-Madison.

I thank the staff of the F.W. Olin Library of my undergraduate alma mater, Drury University, Springfield, Missouri, for the opportunity to use the elegantly handsome Harwood Memorial Reading Room during a number of visits. All libraries are beautiful; Drury's Olin is exceptionally so (as is the Walter Library at the University of Minnesota).

Cindy Ulrich, Carnegie Library of Pittsburgh, helped find the historical location of Fisher Scientific where Jim Miller got his chemicals as a young man. Akiko Ito, Ph.D., Silver Springs, Maryland, helped with information that was available only in the Japanese language. Karen Godshall searched and found a file copy of the 1986 Wisconsin Chapter of the National Organization for Women's (NOW) Calendar and generously made the June illustration of Elizabeth C. Miller available for use in this book.

Michiel Thijssen, Sabine Steenbeek, Wendel Scholma and Arco den Heijer of Koninklijke Brill NV were continuous in their support throughout the publishing effort. I thank them for their support and encouragement.

While I render thanks and appreciation to the readers who took of their valuable time to critically examine and make comments about the manuscript, I make the standard disclaimer that errors that may have crept into the manuscript are mine alone and not those of my helpful readers.

Beverly Kerr McKinnell, my beloved spouse of many years, the mother of my cherished children and the grandmother of my splendid grandchildren, knew of my desire to write this book. She was an ardent feminist and expressed enthusiasm for this endeavor, at least in part, because I was writing about a *woman* scientist (and her husband) and their joint scientific work. Beverly died of cancer too soon to see the fruit of my efforts. I like to think that she would have read this biography with interest and enthusiasm.

Also by Robert G. McKinnell

Cloning: Nuclear Transplantation in Amphibia. Minneapolis: University of Minnesota Press, 1978

Cloning: A Biologist Reports. Minneapolis: University of Minnesota Press, 1979. Published in German as *Cloning: Leben aus der Retorte*. Translated by Roland Bülow. Karlsruhe: G. Braun, 1981

Differentiation and Neoplasia (with others). Berlin, Heidelberg: Springer-Verlag, 1980

Cloning of Frogs, Mice and Other Animals. Minneapolis: University of Minnesota Press, 1985

The Biological Basis of Cancer (with others). Cambridge: Cambridge University Press, 2006. Also published in Japanese by Igaku Shoin, Tokyo, 2000 and in Chinese by Qinghau daxue, Beijing, 2003

Prevention of Cancer. New York: Facts on File, 2008

In Appreciation

A portion of the production cost of this book was made by generous gifts from:

The College of Biological Sciences
University of Minnesota
Saint Paul, Minnesota, U.S.A.

and

An anonymous gift in memory of the
Coolidge/Heinz family
of Beardstown, Illinois, U.S.A.

List of Figures

CHAPTER 1

Cancer is the *Most* Important Disease of the 21st Century: A Reality Most Fail to Comprehend

That *Reality* is Why I Wrote This Book

The loss of a loved one—a child, spouse or a parent—is a bereavement that is beyond description using ordinary words. That loss is neither greater nor less depending upon *how* the death occurs. A child's death that results from being hit by a drunk driver is no more or less heartbreaking than the demise of a child due to a disease. However, the manner of loss may lead survivors to focus on the cause(s) of mortality with the hope of preventing future deaths.[1]

As informed citizens, we all think that heart disease is the number one killer in the United States—certainly many publications affirm this notion.[2] However, the ranking of the numbers of individuals who die of a particular condition, in this instance heart disease, is not necessarily a salient mode for determining significance. Consider the following: compare the loss of life due to cancer of a 5 year old child versus the heart attack death of an 89 year old man. The child has been deprived of approximately 80 years of life expectancy whereas the elderly man has been deprived of essentially no years of life—because he has already exceeded the life expectancy he had at birth.[3]

Person Years of Life Lost

The United States National Cancer Institute has published these starkly compelling words: "*Cancer is responsible for more person years of life lost than all*

1 The fewer losses referred to here would hopefully ensue from enhanced understanding, prevention and control of the cause of death.

2 Xu, J. 2013. Number of deaths from 10 leading causes—National Vital Statistics System, United States, 2010. *JAMA* 309:1582. Heron, M. 2013. Deaths: leading causes of 2010. *National Vital Statistics Reports* 62(6):1–99. Murphy, S.L., J. Xu, K.D. Kochanek. 2013. Deaths: Final data for 2010. *National Vital Statistics Reports* 61(4):1–119.

3 Actually, as long as an individual remains alive, there is life expectancy as yet unfilled. For an 89 year old male citizen of the United States, there are yet about 4 years of life expectancy. These years added to the initial life expectancy ensue because the elderly individual did not die of the multiple causes of death that claimed his compatriots.

 | DOI 10.1163/9789004286801_002

other causes of death…" (Italics added).[4] Figure 1.1 reveals that 8,756,000 person-years of life were lost by Americans in the year 2008 (the most recent data available). Person-years of life lost for *other causes* of death was less in that same year. These other causes include the person-years of life lost due to heart

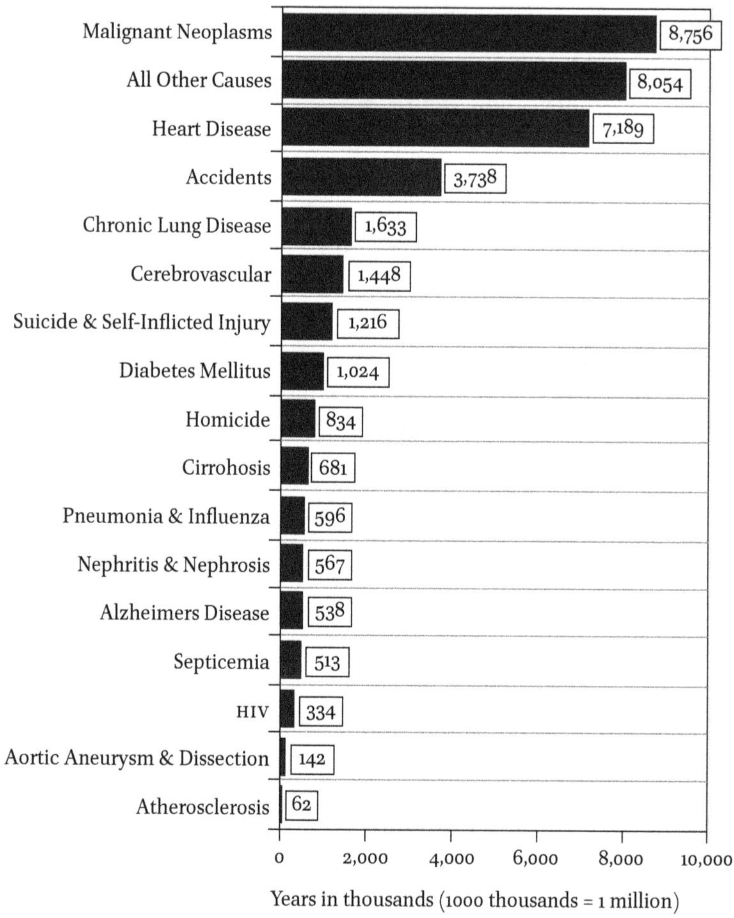

FIGURE 1.1 *Person-years of life lost in the United States due to major causes of death, 2008. Source: National Cancer Institute. Cancer Trends Progress Report—2011/2012 Update, http//progressreport.cancer.gov. 06 March 2015.*

4 National Cancer Institute 2012. Cancer trends progress report—Person-years of life lost. progressreport.cancer.gov/sites/default/files/archive/report2011.pdf. See also: Brustugum, O.T., B. Møller, Å. Helland. 2011. Years of life lost as a measure of cancer burden on a national level. BJC (*British Journal of Cancer*) 111:1014–1020.

disease, chronic lung disease, Alzheimer's disease, suicide, homicide, HIV and the other kinds of death depicted in Figure 1.1. The author of this book is an American which explains his focus on data from the United States. However, the impact of cancer is world-wide.

New Cases and Deaths

Of course, years of life lost is not the only way of indicating the importance of cancer. New cases and deaths by this pernicious malady validates its lethal role in the lives of many humans. There will be an estimated 1,658,370 new cases of cancer with 589,430 deaths in 2015 in the United States.[5] There will be an estimated 14.1 million new cases with 8.2 million deaths worldwide in 2015.[6] In fewer words and numbers, John and Ross[7] report that cancer is the world's leading cause of death.

The global cancer burden is unevenly distributed with a disproportionate impact on high-income regions in North America, Western Europe, Japan, the Republic of Korea, Australia and New Zealand.[8] Not surprisingly, cancer causes more economic loss (productivity) than all of the other causes of death in the world. Readers of this book are admonished to *never consider cancer as a disease second to any*.[9] The overwhelming impact of cancer to human health has been a driving force that led to the writing of this book as I am certain it was in driving the research of Elizabeth C. Miller and her husband James Alexander Miller (Cover illustration and Figure 1.2).

Chemical Carcinogenesis

In chapter two, the multiple etiologies of cancer will be considered. The principal scientists of this book, the Millers, could not study all kinds of cancer.

5 American Cancer Society, Facts and Figures 2015, *Atlanta: American Cancer Society*, 2015.

6 American Cancer Society, Global Facts and Figures, 3rd edition, 2015. *Atlanta: American Cancer Society*.

7 John, R.M., H. Ross. 2010. The global economic cost of cancer. *American Cancer Society, Inc.* publication No. 005444.

8 Forman, D., J. Ferlay. 2014. The global and regional burden of cancer. In: B.W. Stewart, C.P. Wild (eds.), *World Cancer Report 2014*. Lyon: International Agency for Research on Cancer.

9 Obviously, this author is not alone is considering the import of cancer. For instance, see the recent elegant prize-winning account of cancer: Mukherjee, S. 2010. *The Emperor of all Maladies: A Biogaphy of Cancer*. New York: Scribners.

FIGURE 1.2 *James A. Miller and Elizabeth C. Miller.*
SOURCE: NATIONAL CANCER INSTITUTE (U.S.), CLOSING IN ON CANCER: SOLVING A 5000-YEAR-OLD MYSTERY. BETHESDA, MARYLAND: U.S., DEPARTMENT OF HEALTH AND HUMAN SERVICES, PUBLIC HEALTH SERVICE, NATIONAL INSTITUTES OF HEALTH, 1998. N.I.H. PUBLICATION NO. 98-2955.

Hence they focused on "chemical carcinogenesis", i.e., cancer resulting from exposure of vulnerable individuals to certain chemicals in their environment. Environmentally induced cancer has been *grossly underestimated* according to the United States President's Cancer Panel (see discussion in Chapter 3). Was it luck or were the Millers prescient with regard to their recognition of the importance of chemical carcinogenesis? I suggest that it was the latter. The area of research chosen by the Millers, well over a half century ago, has emerged to be of paramount significance. It was likely of comparable importance during those bleak days of World Wars II at the outset of the Millers' endeavors. At that time, the importance of chemical carcinogenesis was only recognized by a few—and these few included the Millers.[10]

10 Elizabeth, in her presidential address to the American Association for Cancer Research, stated that interest in chemical carcinogenesis may be traced to epidemiologists who believed that 60 to 90% of human cancers have important environmental factors in their causation. Miller, E.C. 1978. Some current perspectives on chemical carcinogenesis in the human and experimental animals: presidential address. *Cancer Research* 38:1479–1496.

Two important concepts are associated with environmentally induced cancers: (1) they are exceedingly important forms of cancer due to their great numbers, and (2) identification of specific cancer-causing chemicals and knowledge of how those chemicals impart their malignant mischief may result in eradication of those substances from the environment with a reduction of malignancies. Of course, eradication is not always possible. In those cases where eradication is not feasible, exercising uncompromising control over exposure[11] to the noxious substance(s) serves the same purpose. The Millers' research contributions enormously enhanced understanding of these concepts.

Elizabeth C. and James A. Miller

This volume is thus a biographical account of the Millers who gave to all humankind a timeless understanding of cancer, the *most important disease.* Exceptionally few laypeople have ever heard of the Millers despite their being deeply appreciated by their scientific colleagues.

The associates, students and co-workers of the Millers knew them well and with much affection. During the several years that I worked on this book, I was unable to find anyone who had anything unpleasant or disagreeable to relate about the Millers. Dr. Henry Clement Pitot III (Figure 1.3), Professor of Oncology and Pathology Emeritus and Emeritus Director of the McArdle Laboratory for Cancer Research, University of Wisconsin-Madison and Chair of the President's National Cancer Advisory Board (1976–1982), compared the work of the Millers with that of Edward Jenner (1749–1823) who developed smallpox vaccine. "I put the Miller's discovery in the same category [as the discovery of the vaccine]".[12] Further, Dr. Pitot stated that the research of Elizabeth and James Miller "... *may prove to be the most important of this century in our understanding, prevention, and control of human cancer*" (italics added).[13,14]

---

11 A large chemical company found that their workers may have been exposed to vinyl chloride which was leaking during the production process. The company exerted maximal control by tightening leaks thereby protecting workers and, incidentally increasing the final product PVC because of loss control.

12 Wineke, W.R. October 16, 1987. Obituary, *Wisconsin State Journal.*

13 Memorial tribute of the National Academy of Sciences, undated; see letter to Dr. James A. Miller from Peter H. Raven, Home Secretary, National Academy of Sciences, Washington, D.C., 30 March 1988.

14 Comments made at the Dedication of the Miller Classroom, McArdle Cancer Research Laboratory, Madison, Wisconsin, 19 October 2007.

FIGURE 1.3 *Henry Clement Pitot, III, Emeritus Director of the McArdle Laboratory for Cancer Research, University of Wisconsin-Madison and former Chair of the President's (U.S.) National Cancer Advisory Board.*
COURTESY OF THE UNIVERSITY OF WISCONSIN-MADISON ARCHIVES, IMAGE #S15520.

A century is one hundred years—for something to be the most important in that duration of time is sublime indeed. It is my judgment that Dr. Pitot's words regarding the "understanding, prevention, and control of human cancer" are elegantly suited to describe the story of Elizabeth and James. Thus, with Dr. Pitot's permission, I have used his words as part of the title of this book.[15]

Of course, Dr. Pitot was not alone with his praise of the Millers. The late Dr. Van Rensselaer Potter, (1911–2001), a pioneer bioethicist and cancer researcher at the University of Wisconsin-Madison, in 1987 stated "How sad it is that Jim and Betty [Miller] can never share the Nobel Prize, which they

15 Personal email with permission from Dr. Pitot dated 28 January 2015.

unquestionably deserve, because the prize is given only to living persons."[16] The memorial resolution of the McArdle Laboratory at the death of Elizabeth C. Miller stated that her contributions to the understanding of chemical carcinogenesis are "unrivaled". The comments of Drs. Pitot, Potter and the memorial committee, and others who knew them best, are reason enough to write a biography of this extraordinarily remarkable couple.

The Millers' colleagues praised them without reservation *and* the Millers were similarly recognized within the fraternity of cancer researchers. Scrutiny of Awards and Honors garnered by the Millers (Appendix 1) and a view of the Gallery of *Cancer Research* covers that featured Elizabeth and James (Appendix 3) make it manifestly clear that people familiar with their body of work include it with the very best in the history of medicine. The nexus of chemicals, causes and cancer grew to include the Millers for all time.

A Lack of Recognition

Dr. Henry Pitot commented on the "lack of international recognition" of Elizabeth and James Miller. The lack of "international" recognition actually includes national and local recognition. With this biography I wish (and hope) to diminish that deficiency. It is my opinion that the Millers should be recognized by the myriads of ordinary people whose lives have been impacted for the better.[17]

There is a full-text searchable website devoted to *The History of Women at the University of Wisconsin* (http://digital.library.wisc.edu/1711.dl/UW.UWWomen). Several books by Marian Swoboda and Audrey J. Roberts report on important women at the University of Wisconsin-Madison. Incredible as it may seem, I was unable to find any mention of the extraordinarily distinguished Elizabeth Cavert Miller at this website.[18] Elizabeth's doctoral dissertation

16 Conney, A.H., M.C. Poirier, Y-J. Surh, F.F. Kadlubar. 2009. Elizabeth Cavert Miller (1920–1987) and James A. Miller (1915–2000): A Biographical Memoir. Washington, D.C.: *National Academy Sciences.*

17 The author of this book was cheered to discover a nice photograph of Elizabeth Cavert Miller and James A. Miller, in a discussion of their contributions to metabolic activation, in: Josephy, P.D., B. Mannervik. 2006. *Molecular Toxicology*, 2nd Ed., New York: Oxford University Press, page 519. This reference has been included to indicate that not everyone failed to recognize the Millers—and there were others.

18 I sought confirmation that there was no mention of Elizabeth Cavert Miller's name in this extensive source for Wisconsin women. This was confirmed on 1 February 2007 by a staff member of the University of Wisconsin Digital Collections who searched the files by

adviser, Carl August Baumann,[19] died in 1999. The faculty of the University of Wisconsin-Madison prepared a memorial resolution for Baumann in 2002.[20] In the resolution, it was stated that Professor Baumann had trained over 50 graduate students and many had distinguished careers in science. Three who had joined the faculty at the University of Wisconsin-Madison were named. To my amazement, neither Elizabeth nor her husband Jim were mentioned, despite their long and extraordinarily important careers at the University of Wisconsin-Madison. However, the earlier oversight of the Millers among the distinguished students of Baumann was rectified in the biographical article on Baumann published in 2004.[21]

It is not worth the effort to check *every* biographical compilation to see who has or has not listed the Millers as important contributors to cancer research. However, I agonize as I record here that several recognized lists of distinguished scientists fail to include the Millers. I hasten to add that other biographical compilations do. The deficient works ("deficient" only in the sense that they do not mention the Millers) include Daintith,[22] Millar et al.,[23] Carey,[24] Daintith et al.,[25] Muir,[26] Proffitt,[27] Porter,[28] Grinstein et al.[29] and the 4 volume,

---

keywords and came up with nothing. I checked again in April 2015—still no mention of Elizabeth Cavert Miller.

19 Professor Carl August Baumann is considered much more fully in Chapter 7.

20 Memorial Resolution of the Faculty of the University of Wisconsin-Madison on the Death of Professor Emeritus Carl August Baumann, Faculty Document 1625, 1 April 2002.

21 Suttie, J.W., P.B. Swan. 2004. Carl August Baumann (1906–1999). *Journal of Nutrition* 134:1010–1012.

22 Daintith, J. (ed.), 2009. *Biographical Encyclopedia of Scientists*, 3rd ed. Boca Raton, Florida: CRC Press.

23 Millar, D., I. Millar, J. Millar, M. Millar. 1989. *Chambers Concise Dictionary of Scientists*. Cambridge: Cambridge University Press. Millar, D., I. Millar, J. Millar, M. Millar. 2002. *The Cambridge Dictionary of Scientists*. 2nd ed. Cambridge: Cambridge University Press.

24 Carey Jr., C.W. 2006. *American Scientists*. New York: Facts on File.

25 Daintith, J., D. Gjertsen (eds.), 1999. *Oxford Dictionary of Scientists*. Oxford: Oxford University Press.

26 Muir, H. 1994. *Larousse Dictionary of Scientists*. New York: Larousse Kingfisher Chambers, Inc.

27 Proffitt, P. (ed.), 1999. *Notable Women Scientists*. Detroit: Gale Group.

28 Porter, R. (ed.), 1994. *The Biographical Dictionary of Scientists*. 2nd ed. New York: Oxford University Press.

29 Grinstein, L.S., C.A. Biermann, R.K. Rose. 1997. *Women in the Biological Sciences, A Bibliographic Sourcebook*. Westport, Connecticut: Greenwood Press.

2752 page monumental work of Smith and her colleagues.[30] On a happier note, from my point of view, Ogilvie and Harvey[31] and Bailey 1995[32] (but not Bailey 1998) have informative accounts of Elizabeth's (and, where appropriate, that of James's) life with references. In the award-winning two-volume history of women in science by Margaret W. Rossiter,[33] Elizabeth is cited in Table 5.8, pages 115–120, volume 2 in two fields, *viz*., biochemistry and medical sciences, but here only as *one member of a scientific couple*. Elizabeth and her husband are again referred to in the same Rossiter volume on page 140, with regard to anti-nepotism rules at the University of Wisconsin. It seems to me that the minimal citations to Elizabeth by Rossiter are hardly a ringing endorsement of her (and her husband's) contributions to cancer research. This brief discourse of those who did not know the Millers, or knew them but for some cryptic reason failed to acknowledge that they knew them, mobilized my desire and ambition to write this biography.[34] I never regretted the effort or time that it took.

Who Does Cancer Research?

Early on in my teaching, I found that one member of the two person research team of Elizabeth and James Miller took her undergraduate degree in agricultural biochemistry[35] at the University of Minnesota. That was a fortuitous discovery for my teaching at the University Minnesota, because Elizabeth's undergraduate education could serve to point out that not all distinguished scientists were *men* nor that they obtained their academic credentials from

30 Smith, B.G. (ed.), 2008. *The Oxford Encyclopedia of Women in World History*. Oxford: Oxford University Press.

31 Ogilvie, M., J. Harvey. 2000. *The Biographical Dictionary of Women in Science*. New York: Routledge.

32 Bailey, M.J. 1994. *American Women in Science, A Biographical Dictionary*. Santa Barbara, California: ABC-CLIO. Bailey, M.J. 1998. *American Women in Science 1950 to the Present*. Santa Barbara, California: ABC-CLIO.

33 Rossiter, M.W. 1995 and 1982. *Women Scientists in America*, Volumes 1 and 2, Baltimore: Johns Hopkins Press. Rossiter in an endnote, page 430, number 37, incorrectly identifies her source as "Small" which should be "Smail".

34 Another very personal reason for my writing follows: Elizabeth Cavert Miller cited a study of mine in her 1978 Presidential Address to the American Association for Cancer Research, Inc (*Cancer Research* 38:1479–1496). My ego received a modest boost with her citation—I have not forgotten.

35 The Department of Agricultural Biochemistry was founded in 1917. It is now known as the Department of Biochemistry, Molecular Biology and Biophysics.

prestigious universities such as Harvard, Yale and Princeton in the eastern United States or the University of California, Stanford University or the California Institute of Technology in the west. Elizabeth and her career demonstrated clearly that an extremely bright and dedicated young woman, from a premier public university in the rich heartland of the United States, could rise in academe to make truly significant and lasting scientific contributions.

I never taught a "scientific facts only" cancer course at the University of Minnesota. Rather, in addition to the numerous and often perplexing concepts of the biology of cancer, I presented the lives of the scientists who made significant contributions to cancer research. Who were they, where were they born and where were they educated—what motivated them (if anyone knew) to spend a career devoted to research when many another field paid better with shorter working hours. I wanted to emphasize to students that research is *not* done by universities, colleges, institutions, governments, funding agencies nor committees, but rather by devoted, energetic and brilliant scientists working hard at the laboratory bench.

Other cancer researchers seem equally interested in the people who perform cancer research. Consider what a president of the American Association for Cancer Research, Inc. wrote about this issue: "But the one, single, most important, absolutely indispensable, essential element in this milieu... is the cancer investigator. Without him, there is no effort. Remove him and all the other persons and things associated with cancer research add up to nothing... Motivation, imagination, inspiration, mysticism, contemplation, logic, curiosity, and enthusiasm are among the traits... used by investigators to accomplish their research."[36] It is my wish and hope that other teachers consider these attributes when describing cancer researchers.

Glazed Eyes Not Permitted

I have pondered the "why" of the Millers' lack of name recognition by ordinary people despite their noteworthy contributions to cancer research. I have come to the conclusion that news reporting of a half century ago was really little different from our contemporary 30-second sound bite TV news services of today. Many (most?) reporters are simply incompetent to engage an audience with the description, for example, of the "N-hydroxylation of 2-acetylaminofluorene

36 Stewart, H.L. 1959. The cancer investigator. *Cancer Research* 19:804–818. Note: Stewart was writing over a half century ago. Without doubt, if Stewart were writing today, he would use "he/she" (or "she/he") where "he" occurs in the quotation.

to yield the more carcinogenic N-hydroxy-2-acetylaminofluorene" or other such complex and compelling phrases. A reporter with a glazed-eye audience is hardly what the networks sought then or now. It is my profound hope and wish that eyes are (or become) immune to glazing when reading this book—my reviewers have helped greatly in that respect. Perhaps another reason Elizabeth and James Miller have escaped widespread public recognition was the subject of their research. It is far easier to capture audience attention with the therapeutic power of a new anti-cancer drug, radiation innovation or a novel surgical procedure, in a news release than it is to spark the imagination of lay readers by describing enzymatic biotransformation of certain chemicals. Fortunately, the lack of public acclaim never inhibited the Millers' research.

CHAPTER 2

Cancer for Novices: An Introduction

Understanding cancer has the happy potential of ultimately leading to its prevention and control.[1] Unfortunately, little was known about cancer *causation* prior to the work of Elizabeth C. and James A. Miller, the principal scientists of this book. That statement is particularly true with regard to molecular mechanisms of cancer causation. The reader who is already knowledgeable concerning cancer may wish to skip this chapter. However, the chapter is provided for readers who may wish to learn more about the biology of cancer to enhance their appreciation of the accomplishments of Elizabeth and James Miller. Those accomplishments, as we will see in subsequent chapters, did indeed lead to new modes of prevention and control.

Cancer Defined Sans Torture

The German pathologist Rudolf Virchow[2] (1821–1902) (Figure 2.1) is believed to have said that "I do not think that a living human being can be found that even under torture could actually say what tumors[3] really are."[4] My medical

1 "Understanding" involves much more than the ability to recognize a cancer and predict outcome. Obviously greater understanding of the feared malignancy pancreatic cancer (as an example) will lead to enhanced survival. This is already happening with that particular malignancy—and others. Despite some advances in survival, the reader is urged to note and take seriously the word "potential" in this sentence.

2 Rudolf Ludwig Karl Virchow served as Chair of Pathological Anatomy at the University of Wűrzburg (1849–1856) and Chair of Pathological Anatomy and Physiology at Berlin (1856–1902). Virchow popularized the dictum "*Omnis cellula e cellula*" at a time of uncertainty concerning the origin of cells. He made major contributions to medicine, biology, and archaeology.

3 Biologists and physicians often use the terms "tumor" and "cancer" interchangeably. This may be bewildering to the lay reader. Etymologically "tumor" means a swelling. This can occur with infection or quite naturally by growth as in the breasts of a young woman. Swelling may also ensue from limited cell division forming a *benign* tumor. However, a tumor may be due to uncontrolled cell multiplication and progression to the malignant state. In this latter case, the term "tumor" is synonymous with "neoplasm" and is frequently (often) used synonymously with "cancer" as in the quotation from Virchow.

4 See: Stewart, H.L. 1959. The cancer investigator: Presidential address. *Cancer Research* 19:804–818.

 | DOI 10.1163/9789004286801_003

FIGURE 2.1 *Rudolf Ludwig Carl Virchow (1821–1902), German scientist, is frequently referred to as "the father of modern pathology".*
Source: http://ihm.nlm.nih.gov/images/B25667.

dictionary[5] challenges Virchow by doing exactly that. The dictionary definition includes words such as "invasion and metastasis" and contains the bleak prognosis concerning cancer that "the natural course of which is fatal." The distinctive features that cancer cells hold in common, i.e., the "*hallmarks*" of cancer, have been reviewed by Harahan and Weinberg.[6] The hallmarks comprise

5 *Dorland's Illustrated Medical Dictionary*, 30th ed., 2012, Philadelphia: Saunders (Elsevier).
6 Harahan D., R.A. Weinberg. 2011. Hallmarks of cancer: The next generation. *Cell* 144:646–674.

not a definition (Virchow would be pleased) but rather a detailed description of the several acquired capabilities that cancer cells hold in common. These include (in the words of Harahan and Weinberg) sustaining proliferative signaling, evading growth suppressors, resisting cell death, enabling replicative immortality, inducing angiogenesis, and activating invasion and metastasis. While granting these capabilities do not form a definition, they could well be termed "commonalities" among the diverse cancer cells. My motive in writing this chapter is to provide an *introduction* to cancer biology[7] followed by my thoughts on chemically-caused cancer—the specific research area of Elizabeth and James Miller. Antiquity is a good starting place in an introduction to cancer.

The Antiquity of Malignancy

Humans have long had to cope with this horrific disease as evidenced perhaps by the Ebers Papyrus (Figure 2.2) written about 1,600 BCE.[8] Actually, it is thought that much of the Ebers Papyrus was copied from earlier records dating back 3,000 years BCE—the 3,000 years BCE plus the 2,000 years of the current era give an approximate age of 5,000 years. The reason for using the term "perhaps" in the first sentence of this paragraph is that the meaning of words used 5,000 years ago is, for the most part, lost in time. An Egyptian term translated as "tumor" may refer to benign lumps, swellings or lesions or to a malignant growth. At the very least, the Ebers Papyrus and other ancient Egyptian documents clearly indicate a concern for medicine and surgery thousands of years ago.

A more recent voice from antiquity is that of Hippocrates, ca. 460–ca. 377 BCE, the Father of Medicine. Hippocrates was born on the Greek Island of Cos[9] and had the honor of naming the group of diseases now known as cancer.[10] While little was known about cancer in Cos (and elsewhere) at the

7 Pitot, H.C. 2012. Cancer—An overview. In: R.E. LaFond, (ed.), *Cancer: The Outlaw Cell*, 3rd ed., New York: Oxford University Press, pp. 1–13.

8 Named for Georg Moritz Ebers who obtained the papyrus in 1873 at Luxor. It is now held at the Library of the University of Leipzig.

9 Cos, also known as Kos, is one of the Dodecanese Islands in the southeastern Aegean Sea located closer to Turkey than to Greece. It is mentioned in Acts 21:1 of the Christian Bible.

10 Naming cancer is not the same as understanding cancer. *Understanding* cancer involves knowledge of histology, pathology, biochemistry, genetics, cell biology and much more. Alas, Hippocrates did not have such sophisticated knowledge. Thus, "karkinos", meaning "crab", was used by Hippocratic writers to describe intractable ulcerations as well as noxious growths that seem to correspond to what we now know as cancer. Keil, H. 1950.

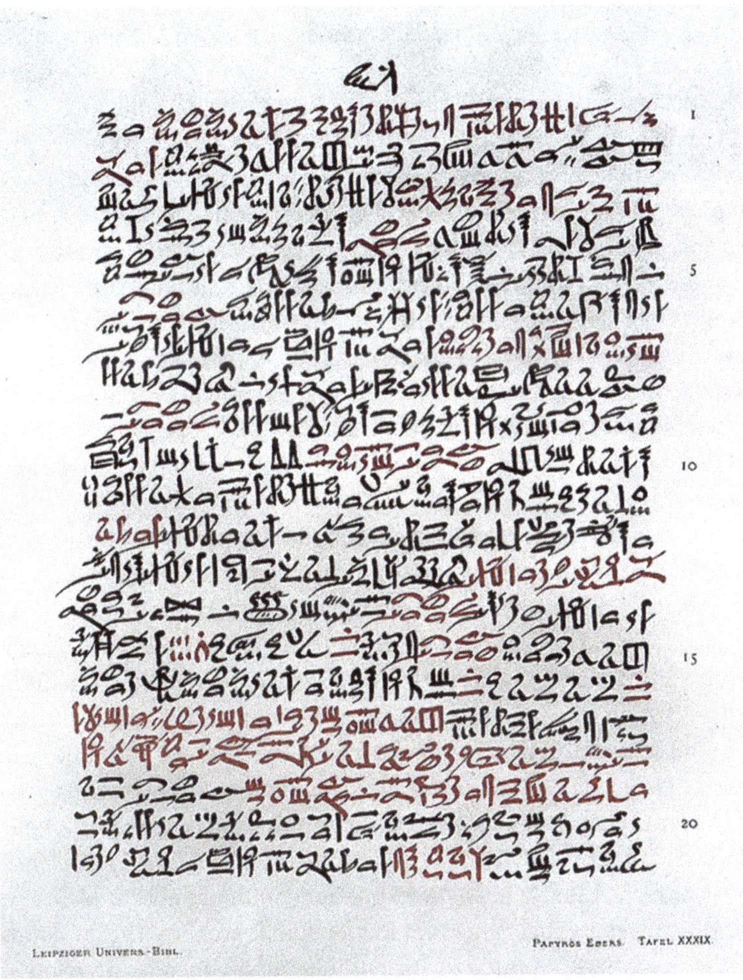

FIGURE 2.2 *One of many "pages", actually a scroll many feet long, of the Ebers papyrus. The papyrus is a medical encyclopedia dating to about 1,500 years* BC.
WELLCOME IMAGES PHOTO #M0008455.

time of Hippocrates, he did indeed recognize the existence of breast, uterine, stomach, skin and mouth cancer. Two aphorisms of Hippocrates are worthy of note here: "First do no harm" and "It is better not to apply any treatment in cases of occult cancer; for, if treated, the patient will die quickly; but if not

The historical relationship between the concept of tumor and the ending –oma, *Bulletin of the History of Medicine* 24:352–377.

treated, they hold out for a long time."[11] Essentially all patients will applaud the first; the second aphorism is significant because it highlights how much has been accomplished in cancer treatment in the past several millennia—and how much remains to be done.

When close to the surface, a superficial cancer nodule could be gently pushed or displaced. When this happened, the nodule often seemed to be anchored in the tissue, and columns of something (cells) seemed to radiate from the nodule. It has been said that those multiple columns of cells reminded Hippocrates and his students of the limbs of a crab. In contrast, benign tumors tended to be encapsulated, and thus the limb-like radiating columns of cells that seem to anchor malignant tumors do not exist in the benign growths.[12] The scientific name of the edible crab is *Cancer*, the equivalent of the Greek word *Karkinos*. The contemporary term "carcinoma"[13] is obviously derived from the Greek and the term persists to this day.

Of course, cancer as a disease predates the Egyptians at Luxor and Hippocrates of Cos by many thousands of years. Some human fossils display evidences of cancer. However, there are many problems with the detection and diagnosis of cancer in these ancient remains. One problem relates to a paucity of cases. For the most part cancer is a disease of older people. Humans in prehistory had shorter life expectancies than we enjoy today. Consequently, it would be anticipated that evidence for the disease would be found at a reduced prevalence in fossils and this is the case. Another factor related to the dearth of cancer in human fossils is the absence of industrial chemicals, toxic wastes, and vehicular exhaust fumes at that time. Couple a lower prevalence of cancer with the fact that human *fossils* are not common to begin with, and there emerges the reality that cancers of human fossils are rare. The few kinds of cancer that have been reported in ancient humans are mostly those found in bone, for the obvious reason that bones are generally all that remains of an ancient human. Examples human cancers found (detected) in fossils include multiple myeloma, a cancer that leaves distinctive lesions in the bones of afflicted individuals, and melanoma, detected by pigmented foci found in the skull of fossil humans. While not in bony tissue, ovarian and uterine cancers

11 Shimkin, M.B. 1977. *Contrary to Nature*. Bethesda, Maryland: U.S. Department of Health, Education and Welfare, Public Health Service, National Institutes of Health (NIH Publication 76–720).

12 It is the business of the physician specialist to diagnose cancer which is confirmed by a pathologist—the illustration given here of the appearance of superficial nodules is just that, an illustration, and readers are strongly cautioned against self-diagnosis.

13 Carcinoma is defined as a cancer originating from epithelial cells.

have been discovered in mummies of Egyptians and Pre-Columbian Incas, as well as in other fossils dating to thousands of years ago.[14]

A difficulty of working with fossils and their tumors is exemplified by the Kanum jaw tumor discovered in 1932 by Louis S.B. Leakey in the east African nation of Kenya.[15] The jaw is approximately a million years old and if its abnormal growth is in fact malignant, the "cancer" would be one of the earliest yet discovered. But is the million year old "cancer" a human cancer? The jaw of the early hominid is clearly *not* that of a modern human but is it sufficiently "human" to be included in this discussion? I would judge probably "yes". The tumor has been thought to be a possible early example of Burkitt's lymphoma.[16] The lymphoma afflicts primarily jaws of young individuals. The Kanum fossil is that of a jaw of a young person. However, because the interpretation of archaic remains is difficult indeed, much of the biology and pathology of the Kanum jaw remains enigmatic.

Let us move from ancient bones to a more contemporary view of cancer. A simplistic distinction of benign tumors from malignant tumors is that the latter have the competence to undergo metastasis. That follows.

Metastasis

The term "metastasis" was applied to a form of cancer by Joseph Claude Anselme Récamier (1774–1852) (Figure 2.3), a physician at the Hôtel Dieu in Paris, in his two volume text on cancer published in 1829.[17] In the book, he

---

14 Weiss, L. 2000. The antiquity of cancer and metastasis. *Cancer and Metastasis Reviews* 19:193–204; Capasso, L.L. 2005. Antiquity of cancer. *International Journal of Cancer* 113:2–13. See also: David, A.R., M.R. Zimmerman. 2010. Cancer is an old disease, a new disease or something in between. *Nature Reviews Cancer* 10:728–733. Binder, M., C. Roberts, N. Spencer, D. Antoine, C. Cartwright. 2014. On the antiquity of cancer: Evidence for metastatic carcinoma in a young man from ancient Nubia (c. 1200 BC). PLoS One Mar 17;9(3):e90924. doi:10:1371/journal.pone.0090924.eCollection. Recently leukemia was believed detected in a 7,000 year old skeleton; see Dr. Heike Scherf, Senckenberg Research Institute and Natural History Museum.

15 Tobias, P.V. 1960. The Kanum jaw. *Nature* 185:946–947. Sandison, A.T. 1975. Kanum mandible's tumor. *The Lancet* 1:279. Stathopoulos, G. 1975. Kanum mandible's tumour. *The Lancet* 1:165.

16 Molyneux, E.M., R. Rochford, B. Griffin, R. Newton, G. Jackson, G. Menon, C.J. Harrison, T. Israels, S. Bailey. 2012. *Burkitt's lymphoma*. The Lancet 379:1234–1244.

17 Récamier, J.C.A. 1829. *Recherches sur Le Traitement du cancer*, 2 vols. Gabon, Paris. See also: Carr, J., I. Carr. 2005. The origin of cancer metastasis. *Canadian Bulletin of Medical*

FIGURE 2.3 *Joseph Claude Anselme Récamier (1774–1852) was a pioneer in the study of cancer metastasis.*
IMAGE C. 1894.

wrote: "Le Fait de M. Parent, en montrant la resolution spontanée d'un autre engorgement de méme nature, peut conduire a admettre des *metastases* cancéreuses" (italics added). Récamier recognized that a breast cancer, growing in the brain, was identical to a primary breast cancer. The breast cancer had obviously spread to the brain. Lacking a microscope he could not state unambiguously how the cells moved from the breast to the brain.[18]

History 22:353–358. The term "metastasis" existed about 200 hundred years prior to the birth of Récamier. However, Récamier is given credit for its use in pathology to describe a cancer growing at a site distant from the primary tumor.

18 Lugassy, C., J.-P. Escande. 1997. The haematogenous theory of metastasis: Récamier did not propose it. *Virchows Archiv* 431:371.

We now know that cancer cells have the competence to disseminate via the lymphovascular[19] system, or by other means, and grow as colonies at inappropriate sites throughout much of the body. Many consider metastasis to be *the* defining attribute of cancer.[20] Of course, for metastasis to occur, a whole series of events must happen in such a manner that some cells from primary the malignancy grow at a distant site. The likelihood of any single cancer cell producing a metastatic colony is believed to be infinitesimal. It is only because of the enormous number of cancer cells released by the primary tumor that metastasis is successful, when it is indeed successful.[21,22]

Chemotherapy

Cancer cells resemble normal cells in many ways, because malignant cells *are* progeny of the body's normal cells. Cancer cells *are not* enemies from without.

---

19 The adjective "lymphovascular" is not in my 2006 medical dictionary nor any other dictionary I have scrutinized. I use it to save space, as have many other scientists. Like normal tissues and organs, cancers require a nourishing blood supply and have lymph vessels which carry tissue fluids and lymphocytes and other cells. The lymph is ultimately returned to the blood. Lymphovascular is a term that includes both blood vessels and lymph vessels. Cancer cells may be spread by either or both kinds of vessels.

20 Examples of this view: Ewing believed that metastasis is the "most convincing evidence" of lawless growth, i.e., cancer. Ewing, J. 1916. Pathological aspects of some problems of experimental cancer research. *Journal of Cancer Research* 1:71–86. "The detection of metastases constitutes decisive evidence for categorizing a proliferating primary lesion, previously of uncertain potential, as neoplastic and 'malignant', and the phenomenon is a topic of unparalleled importance in cancer medicine and biology." Tarin, D. 1992. Tumour metastasis. In: J.O'D. McGee, P.G. Isaacson, N.A. Wright (Eds.), *Oxford Textbook of Pathology*, pp. 607–633. Oxford: Oxford University Press. Metastases "form an ineluctable hindrance to successful therapy". Ochsner, A., M. DeBakey, 1942. Significance of metastasis in primary carcinoma of the lung. *Journal of Thoracic Surgery* 11:357–387. Tarin, D. 2011. Cell and tissue interactions in carcinogenesis and metastasis and their clinical significance. *Seminars in Cancer Biology* 21:72–82. Fidler, I.J., M.L. Kripke. 2012. The biology of cancer metastasis. In: *Cancer: The Outlaw Cell*. 3rd Ed., Oxford University Press, New York, pp. 198–210. The National Cancer Institute sponsors clinical trials with many new cancer treatments under study. These studies include treatment for metastasis.

21 A sobering statistic is that most, perhaps as many as 80%, of cancers are metastatic at the time of diagnosis.

22 The complex nature of metastasis is discussed in: Chaffer, C.L., R.L. Weinberg. 2011. A perspective on cancer cell metastasis. *Science* 331:1559–1564. Talmadge, J.E., I.J. Fidler. 2010. The biology of cancer metastasis: Historical perspective. Cancer Research 70:5649–5669.

Cancer cells *are* autochthonous.[23] And herein lies the problem. The resemblance of cancer cells to normal cells is a major problem in the design of cancer chemotherapeutic drugs. Why? Consider the design of cancer treatment that focuses on cell division. Cancer cannot undergo growth and metastasis without cell division. Hence, cell division is a likely and reasonable target for a cancer drug. Let us postulate a drug that absolutely stops cancer cells dead in their tracks by blocking cell division with a specific DNA inhibitor (a drug that interferes with DNA replication such that cell division becomes impossible—in fact, many such drugs have been devised, including a number of alkylating agents).[24]

There are pitfalls with such drugs. Because normal cells are so similar to their cancer cell progeny, drugs that interfere with DNA replication in cancer cells also interfere with DNA replication in normal cells. Recall from high school biology that many normal cells continue to replicate throughout life. These replicating cells include cells of the intestinal lining, bone marrow and hair follicles. The well known toxicity of chemotherapy is directly related to damage to these *normal* cell types. That damage, which includes nausea, bleeding from the gums and other places, anemia and loss of hair, is directly attributable to the similarity of *normal* cells to cancer cells (or more appropriately, *vice versa*). This brief section on drug design is inserted to augment the notion of the complexity of cancer. Many scientists hope that emerging means of cancer prevention will someday negate the need for chemotherapeutic agents.[25]

Radiation, Viruses, Bacteria, Heredity and/or Possibly "Bad Luck"

A carcinogen is something that has the competence to cause cancer, and the process of cancer induction is referred to as carcinogenesis. Cancer has a variety of causes as indicated by the heading of this section.

---

23 Autochthonous refers to origins and in this instance, the origin referred to is self; i.e., cancer cells originate from the individual who suffers from the cancer.

24 Alkylating agents contain electrophilic alkyl groups that have the competence to covalently bind with cell DNA nucleophilic sites. The binding of the drug with the cell DNA interferes with proliferation of rapidly dividing cancer cells. Some drugs included among the alkylating agents are alkyl sulfonates, nitrogen mustards and nitrosoureas. See Tew, K.D. 2008. Alkylating agents. In: *Cancer Principles & Practice of Oncology*, 8th ed. V.T. DeVita. T.S. Lawrence, S.A. Rosenberg (Eds.), Philadelphia: Wolters Kluwer.

25 For an account of drug design, see: Neidle, S. (ed.). 2008. *Cancer Drug Design and Discovery*. Boston, Academic Press.

FIGURE 2.4 *Wilhelm Conrad Röntgen (1845–1923) received the Nobel Prize for his development of X-rays.*
IMAGE FROM THE HISTORY OF MEDICINE (IHM), U.S. NATIONAL LIBRARY OF MEDICINE, B021997.

Radiation can cause cancer.[26] X-rays were discovered by the German physics professor Wilhelm Conrad Röntgen[27] (Figure 2.4), and it was soon discovered that x-rays under certain circumstances, could induce cancer.[28] Skin cancer

26 Little, J.B. 2000. Radiation carcinogenesis. *Carcinogenesis* 21:397–404.

27 Röntgen, W.C. 1895. Ueber eine neue Art von Strahlen. *Sitzungsberichte Physikalisch-medizinische Gesellschaft zu Würzburg.* 9:132–141. Röntgen was raised in the Netherlands, briefly attended the University of Utrecht, and subsequently obtained a Ph.D. in physics from the University of Zurich (Switzerland). He spent most of his professional life at the University of Munich. He won the first ever Nobel Prize in physics in 1901 for his discovery of X-rays.

28 Henshaw, P.S., J.W. Hawkins. 1944. Incidence of leukemia in physicians. *Journal of the National Cancer Institute* 4:339–346.

is related to excessive exposure to radiation of the sun, more specifically, to that kind of radiation known as ultraviolet B or UVB.[29] Skin cancers are the most common malignancies in the United States with about 3.5 million cases of basal cell and squamous cell (non-melanoma) skin cancers, occurring annually. Malignant melanoma skin cancer is far more lethal and this form of skin cancer has an expected 76,690 cases with 9,480 deaths in 2013.[30,31]

Certain infections by viruses and bacteria are associated with cancer causation.[32] The first cancer known to be caused by a herpesvirus was the kidney cancer of the North American Leopard Frog, *Rana pipiens*.[33] But of course, more than just frogs suffer from herpesvirus malignancies. Human cancers associated with herpesviruses, or thought to have a herpesvirus etiology, include Kaposi's sarcoma, Burkitt's lymphoma, Hodgkin's lymphoma, leiomyosarcoma, B cell lymphoproliferative disorders and nasopharyngeal carcinoma.[34] Much cervical cancer is due to infection by viruses of another group, namely the human papilloma viruses (HPV). Gastric adenocarcinoma and a stomach non-Hodgkin's lymphoma are thought to be caused by, or related to, an infection of the bacterium *Helicobacter pylori*.[35]

---

29 Robinson, J.K., D.S. Rigel, R.A. Amonette. 1997. Trends in sun exposure knowledge, attitudes, and behaviors: 1986 to 1996. *Journal of the American Academy of Dermatology* 37:179–186.

30 American Cancer Society. *Cancer Facts & Figures 2013*. Atlanta: American Cancer Society.

31 Rogers, H.W., M.A. Weinstock, A.R. Harris, M.R. Hinckley, S.R. Feldman, A.B. Fleischer, B.M. Coldiron. 2010. Incidence estimate of nonmelanoma skin cancer in the United States, 2006. *Archives of Dermatology* 146:283–287.

32 Gross, L. 1983. *Oncogenic Viruses*, 3rd Edition. New York: Pergamon Press; Hoppe-Seyler, F., K. Butz. 2002. Viral mechanisms of human carcinogenesis. In: W.B. Coleman, G.J. Tsongalis (eds.), *The Molecular Basis of Human Cancer*. Totowa, New Jersey: Humana Press. pp. 233–247.

33 McKinnell, R.G. 1973. The Lucké frog kidney tumor and its herpesvirus. *American Zoologist* 13:97–114. Genomic studies of the tumor's herpesvirus DNA are relatively more recent, see: Davison, A.J., C. Cunningham, W. Sauerbier, R.G. McKinnell. 2006. Genomic sequences of two frog herpesviruses. *Journal of General Virology* 87:3509–3514. The species with the herpesvirus tumor, *Rana pipiens*, has recently been re-named *Lithobates pipiens*.

34 Buchschacher Jr., G.L., F. Wong-Staal. 2008. Etiology of Cancer: Viruses. In: V.T. DeVita, S. Hellman, S.A. Rosenberg, (eds.), *Cancer: Principles and Practice of Oncology, 8th ed.* Philadelphia: Lippincott Williams & Wilkins, pp. 169–188. Liao, J.B. 2006. Viruses and human cancer. *Yale Journal of Biology and Medicine* 79:115–122.

35 Parsonnet, J., S. Hansen, L. Rodriguez, A.B. Gelb, R.A. Warnke, E. Jellum, N. Orentreich, J.H. Vogelman, G.D. Friedman. 1994. Helicobacter pylori infection and gastric lymphoma. *The New England Journal of Medicine* 330:1267–1271.

Hereditary cancers are believed to be relatively uncommon.[36] Twin studies support the belief that cancer resulting from hereditary factors are indeed numerically of less importance than other cancers. Retinoblastoma and Wilms' tumor (nephroblastoma) are pediatric hereditary cancers. Retinoblastoma is a cancer of an embryonic stem cell which, if untreated, moves from the eye via the optic nerve to the brain and subsequently to other organs, eventually causing death. Wilms' tumor has an origin from malignant kidney stem cells. In addition to the childhood cancers mentioned, a small fraction of prostate, colorectal, and breast cancers have a hereditary component.

"Bad luck" indeed may be a description of what happens to *all* those who are diagnosed with cancer. However, the term is used more specificity to refer to cancer causation and, if a contemporary study of Tomasetti and Voglestein proves to be correct, the "bad luck" factor may be great indeed.[37] The luck factor relates to the relative abundance of stem cells in a particular tissue and their role in cancer as the result of random mutations. Fully differentiated cells do not have cell progeny, i.e., they are post-mitotic. Random mutations have little, if any effect, to post-mitotic cells. However, stem cells continue to give rise to more stem cells and other cells committed to a differentiation pathway. The stem cells continuing mitotic activity thus permits random mutations to be passed on to many cell descendants. Hence, the greater abundance of stem cells, the greater the risk for cancer in those tissues. The "luck" factor relates to the fact that most mutations are random. These studies are recent and further developments are anticipated.

Factors that Give Rise to Cancer: Environment

Environmental factors and/or lifestyle account for more than two thirds of all cancer in the United States, and they together emerge as a major cause of cancer.[38] About 30% of cancer is attributable to tobacco and another 35% is

36 Foulkes, W.D. 2008. Inherited susceptibility to common cancers. *New England Journal of Medicine* 359:2143–2153; see also: Boman, B.M., L. Melchert, J.Z. Fields. 2002. Hereditary cancer. In: W.B. Coleman, G.J. Tsongalis (eds.), *The Molecular Basis of Cancer*. Totowa, New Jersey: Humana Press, pp. 159–185.

37 Tomacetti, C., B. Vogelstein. 2015. Variations in cancer risk among tissues can be explained by the number of stem cell divisions. *Science* 347:78–81. See also: Couin-Frankel, J. 2015. The bad luck of cancer. *Science* 347:12. Wadarz, D., A.G. Zauber. 2015. Risk factors and random chances. *Nature* 517:563–564.

38 Lichtenstein, P., N.V. Holm, P.K. Verkasalo, A. Iliadou, J. Kaprio, M. Koskenvuo, E. Pukkala, A. Skytthe, K. Hemminki. 2000. Environmental and heritable factors in the causation of cancer. *The New England Journal of Medicine* 343:78–85.

due to diet.[39] Other causes are though to be relatively minor, e.g., occupational cancers 4%, cancers due to pollution 2%, and cancers due to reproduction/sexual behavior 7%. Added to this list is cancer due to infectious agents, obesity and inactivity, and a host of other causes—all of which cause chemical changes in cells. No wonder the Elizabeth and James Miller were focused on chemistry and cancer.

Most cancer resulting from exposure to chemicals is environmental. Cancer prevalence varies from country to country by as much as (or more than) tenfold. An example: colon cancer is infrequent among citizens of Ibadan in southwest Nigeria. In contrast, colon cancer prevalence in Connecticut is far higher, as it is in most "developed" western countries. Is this difference environmental or is it due to genetics? Studies of colon, stomach, and breast cancer (as well as most other cancers) reveal the striking differences in prevalence between countries. What is noteworthy is that migrants, and especially the children of migrants, adopt the cancer prevalence of their new country of residence. What has changed is not their heredity but their environment. This has led epidemiologists to assert that much cancer is due to the environment and this in turn suggests that much cancer is preventable. The Millers phrased it this way: "The concept has arisen that certain chemicals in the environment—both man-made and naturally occurring—in food, water, air, drugs, and our intestinal contents or from societal habits—may be important factors in the etiologies of human cancers."[40] That quotation concerning cancer causation of more than a third of a century ago remains scientifically true and still timely.

The concept of chemicals causing cancer grew out of what is now called epidemiology and had its origin well over 200 years ago. Epidemiology relates the vulnerability to a specific disease with exposure to an agent(s) as a result of occupation, lifestyle or place of residence. This has also been termed the natural history of cancer: the geographic distribution, frequency and the circumstances with which cancer develops. This early carcinogenesis research was not experimentally based—it was based on astute observations by great physicians in the past quarter of a millennium. Using more contemporary phraseology, one could say that cancer-causing chemicals were identified by "body-count". Contemporary scientists seek to identify chemicals that may be cancer-causing *prior* to the expression of their body-count potential.

---

39 Doll, R., R. Peto. 1981. The causes of cancer: Quantitative estimates of avoidable risks of cancer in the United States today. *Journal of the National Cancer Institute* 66:1191–1308.

40 Miller, E.C., J.A. Miller. 1981. Searches for ultimate chemical carcinogens and their reactions with cellular macromolecules. *Cancer* 47:2327–2345.

FIGURE 2.5 *John Hill, an English physician, wrote on cancers observed in individuals who used snuff.*
IMAGE FROM THE HISTORY OF MEDICINE (IHM), U.S. NATIONAL LIBRARY OF MEDICINE, B010498.

Consider the following examples of early observations relating chemical exposure to cancer. These are the types of observations referred to above as "body-counts". An English physician, John Hill (1716–1775) (Figure 2.5), noted in 1761 that users of tobacco snuff were particularly liable to cancer of the nose.[41] He described a number of cases of nasal growths afflicting snuff takers. In one case, the early and painless nodule grew into a lesion with "all the frightful symptoms" of cancer. He warned against the use of this form of tobacco: "the miserable consequences of indulging in this custom (of using snuff) are plain: and happily the remedy is as obvious: there needs no medicine to combat the

41 Hill, J. 1761. *Cautions Against the immoderate Use of Snuff.* London: R. Baldwin and J. Jackson. John Hill wrote on a remarkable assortment of subjects; some scientific and others not so scientific. He had a medical degree from the University of Saint Andrews, Scotland. He referred to himself as "Sir John" despite King George II's non-fulfillment of the prerequisite touch of the sword with dubbing.

effects of Snuff, nor is any of power to do it; the sole cure and the certain cure is to leave off the custom". If Hill meant by "leave off the custom" either to quit or to never start, his admonishment of a quarter of a millennium ago is clearly absolutely appropriate today. Hill's essay on snuff, credited by Redmond[42] as the first record to suggest an association of tobacco with cancer, went through three editions—obviously it was read. Some years later, Samuel Thomas von Soemmering[43] (1755–1830) reported cancer of the lip thought to be caused by pipe smoking. Note that John Hill's concern for the carcinogens of tobacco predated by 200 years the British report on smoking and health[44] as well as the United State's Surgeon General's Report in 1964.[45] Despite the hoary history of hideous consequences, consumption of tobacco continues. In 1967 two distinguished cancer researchers wrote, "30,000 cases of a practically incurable neoplastic disease could be eliminated if we were wise enough to discard an inane habit."[46] Only the numbers have changed. Deaths in the United States due to smoking are now more than 400,000 per year,[47] about 500% greater than in 1967.

A little more than a decade after John Hill's snuff warning, a surgeon of London's Saint Bartholomew's Hospital, Percivall Pott (1714–1788), (Figure 2.6) wrote that a mixture of noxious substances (i.e., chimney soot), when not removed from the skin after occupational exposure, could cause a specific cancer.[48] Pott reported that young chimney sweeps, also known as "climbing boys", with chronic exposure to soot developed scrotal cancer, which is now

---

42 Redmond Jr., D.E. 1970. Tobacco and cancer: The first clinical report, 1761. *New England Journal of Medicine* 282:18–23.

43 Soemmering, S.T., von. 1795. *De Morbis Vasorum Absorbentium Corporis Humani. Verrentrappii et Wenneri, Trajecti ad Moenum*, Frankfurt, Main. P. 109.

44 *Smoking and Health: Summary and report of the Royal College of Physicians of London on smoking in relation to cancer of the lung and other diseases*. 1962. London: Pitman.

45 *Smoking and Health: Report of the Advisory Committee to the Surgeon General of the Public Health Service*. 1964. US Department of Health, Education and Welfare, Public Health Service, Washington, DC.

46 Shimkin, M.B., V.A. Triolo. 1969. History of chemical carcinogenesis: Some prospective remarks. *Progress in Experimental Tumor Research* 11:1–20.

47 The CDC (Centers for Disease Control and Prevention) estimates that 443,000 deaths occur annually due to cigarette smoking and second hand smoke, data for the United States, 2000–2004.

48 Pott, P. 1775. *Chirugical observations relative to the cataract, the polypus of the nose, the cancer of the scrotum, the different kinds of ruptures and the mortification of the toes and feet*. London: Hawkes, Clarke, and Collins. See also: Potter, M. 1963. Percivall Pott's contribution to cancer research (including a reproduction of Pott's original article). *National Cancer Institute Monograph* 10:1–13.

FIGURE 2.6 *Percivall Pott, a London surgeon who noted that boys who were chimney sweeps, where liable to scrotal cancer. Derived from a portrait by Nathanial Dance-Holland, London, 1785.* IMAGE FROM THE HISTORY OF MEDICINE (IHM), NATIONAL LIBRARY OF MEDICINE #B026992.

known to be a squamous cell carcinoma.[49] The scrotum is the pouch of skin that contains the testes. Pott observed that scrotal cancer was relatively common in young chimney sweeps but exceedingly uncommon among other males. He correctly inferred that the cancer resulted from the retention of chimney soot in the folds of the scrotal skin. While Pott is credited for discovering the etiology of scrotal cancer among chimney sweeps, it is refreshing to note that chimney sweeps with little or no education had already surmised a

49 Shimkin, M.B., V.A. Triolo. 1969. History of Chemical Carcinogenesis: Some Prospective Remarks. *Progress in Experimental Tumor Research* 11:1–20.

relationship: the climbing boys called the cancer the "soot wart". Pott referred to this chimney sweep cancer as a "most noisome, painful, and fatal disease".

Identification of a cancer-causing substance provides an opportunity for controlling contact with or eliminating that substance from the environment. Bathing provides a method for the elimination of soot from the body surface (skin). Upon learning of Percivall Pott's study, bathing became common among Danish chimney sweeps with the salubrious corollary of the virtual disappearance of scrotal cancer. English chimney sweeps, unlike their Danish counterparts, were not motivated to bathe—or lacked the necessary facilities—with the result of a no diminution of scrotal cancer prevalence. Interest in Pott's chimney sweep cancer derived from relating a specific cancer to a specific condition experienced by a particular occupational group with the potential of cancer prevention. Many historians consider Percivall Pott to be the Father of Chemical Carcinogenesis.

About 100 years after Pott's report, tar and paraffin oils were thought to cause skin cancer in workers in the tar and paraffin industry of Halle (a German city between Berlin and Frankfurt). In 1875, Richard von Volkmann,[50] professor of surgery at Halle, described three paraffin oil and coal tar workers with scrotal cancer.[51] Shortly thereafter in 1876, "paraffin workers' cancer" afflicting the scrotum in two Scottish workers was described by Joseph Bell.[52] The workers were exposed to "fumes of paraffin in shale works." Alexander Scott reported that parts of Scotland had "practically unlimited" shalefields.[53,54] Workers of oil shale were vulnerable to cancers which appeared within twenty or more years of work, first as a wart or papule, followed by cancer. Scott deplored the lack of protective clothing and failing that, he urged cleanliness. As noted earlier when considering chimney sweep scrotal cancer, washing was not a high

50 von Volkmann wrote poems and folk tales for children under the *nom de plume* of Richard Leander.

51 Volkmann, R. von. 1875. *Beiträge zur Chirurgie*, Leipzig (cited by Haddow and Kon, 1947, *British Medical Bulletin* 4:314–326). See also von Volkmann, R. 1874. *Ueber Theer- und Russkrebs*. Berliner Klinische Wochenschrift 11:218.

52 Bell, J. 1876. Paraffin epithelioma of the scrotum. *Edinburgh Medical Journal* 22:135–137.

53 Oil shale is a kind of rock with a high content of organic matter. The rock is mined much as is coal, and the broken rock is heated and processed to yield crude oil, which in turn can be distilled and refined into candle wax, kerosene, various lubricants and solvents, gasoline, fuel oil, etc. Scotland mined oil shale starting around 1850.

54 Scott, A. 1923. The occupation dermatoses of the paraffin workers of the Scottish shale oil industry, with a description of the system adopted and the results obtained at the periodic examinations of these workmen. In: *Eighth Scientific Report on the Investigations of the Imperial Cancer Research Fund*. London: Taylor and Francis. pp. 85–142.

FIGURE 2.7 *Ludwig Wilhelm Carl Rehn reported urinary bladder cancer among chemical dye workers.*
IMAGE FROM THE HISTORY OF MEDICINE (IHM), U.S. NATIONAL LIBRARY OF MEDICINE, B08406.

priority among workers. We now know that the offending chemicals in tar and paraffin oils are polycyclic aromatic hydrocarbons (PAH).

"Aniline" cancer, having afflicted chemical dye workers in the production of fuchsin,[55] was reported in 1895 by the German physician Ludwig Wilhelm Carl Rehn (1849–1919) (Figure 2.7).[56] Workers absorbed noxious aromatic amines through the skin and lungs, after which the amines were excreted in the urine.

55 Fuchsin is an aniline dye with a deep red color. The name is derived from the German equivalent "Fuchs" (fox) of the French *Renard*, which was the name of the chemical company that first made the red dye.

56 Rehn, L. 1895. Blasengeschwülste bei Fuchsin-Arbeitern. *Archiv für Klinische Chirurgie* 50:588–600.

Cancer of the urinary bladder followed. "Aniline"[57] cancers were soon reported in other industrialized countries including Switzerland and England. Aniline is an oily organic chemical derived from coal-tar, which is itself a product of the conversion of coal to coke. Aniline is used in the manufacture of a bewildering array of substances, including sulfa drugs, some rocket fuels, pesticides, plastics, and coal-tar dyes. One coal-tar dye with a blue color completely wiped out the historic source, India, of the natural and widely-used blue dye indigo. After World War I, virtually all aniline (and hence aniline dyes) was produced by the chemical industries of Germany. Producing aniline dyes in large quantities resulted in the toxic exposure of industrial workers to certain aromatic amines including ß-naphthylamine, benzidine and 4-aminobiphenyl.[58]

An analysis of 5,142 cases of "occupational cancer" was made much later, after the above studies; skin cancer was found among workers who toiled with pitch, tar, tar-products, shale oil, mineral oil or bitumen.[59] The history of chemical carcinogenesis among industrial workers had the major shortcoming of quantifying diseased and dying industrial workers (that is, counting human fatalities, the "body-counts" referred to previously). Had young chimney sweeps not developed scrotal cancer, there would be no notion that the soot of chimneys was potentially hazardous. The same can be said about pitch and tar workers. Many workers paid with their lives for this information. Would it not be more humane, more merciful, more gentle if there was some way of learning about the carcinogenicity of chemicals without the need of putting people at risk? The Millers must have pondered these questions many times through the years.

In order to learn how chemicals caused human cancers, animal studies were initiated. Certainly a "red flag" would be waving if a chemical substance could be shown to cause cancer in an experimental animal. The red flag might very well have the salubrious effect of saving the lives of many workers. There was yet another reason that animal studies were sought. That reason addressed the concern that most industrial hazard resulted from an exposure to a mix of

---

57 The quotation marks are placed around "aniline" because it is now believed that the bladder cancer-causing agents are aromatic amines, especially 2-naphthylamine, and not specifically aniline, see: Lijinsky, W. 1979. Current status of experimental chemical carcinogenesis and its implications to human cancer risk. *Cancer Research* 39:2887–2890; see also Miller, J.A. 1994. Brief history of chemical carcinogenesis. *Cancer Letters* 83:9–14.

58 Amines are organic compounds derived from ammonia. Aromatic amines have one or more hydrogen atoms replaced with benzene rings or heterocyclic rings.

59 Henry, S. 1947. Occupational cutaneous cancer attributable to certain chemicals in industry. *British Medical Bulletin* 4:389–401.

chemicals. Tobacco smoke, snuff, soot, shale oil, coal-tar and the myriad other toxic material encountered in manufacturing processes are clearly composed of mixtures of chemicals. It became important to know *which* chemical(s) in the mixture caused cancer. Identifying hazardous chemicals provides the intellectual framework on which to devise a protective strategy for workers. However, knowing which chemical was the noxious one was dependent not only on the use of experimental animals[60] but the emergence of procedures that could separate and purify chemicals from a complex mixture; the arrival of animal testing coupled with modern chemistry marked the beginning of experimental cancer research.

Animal Experimentation

A method of producing experimental cancers in animals was introduced by two Japanese investigators, Katsusaburo Yamagiwa (1863–1930) and Koichi Ichikawa (1888–1948) of the Tokyo Imperial University Medical School[61] where Yamagiwa (Figure 2.8) was Director of the Department of Pathology. In the first two decades of the twentieth century, Yamagiwa and Ichikawa will be remembered in cancer history as the first investigators to produce cancer by the application of a chemical agent to an experimental animal.

Yamagiwa and Ichikawa repeatedly painted crude coal tar dissolved in benzene onto the ears of rabbits for many months.[62] They probably chose coal tar because it had previously been shown to cause cancer in humans. Cancers (epithelial carcinomata) were observed in about a third of the 22 surviving rabbits. Lymph node swellings, thought to be metastases, were found in four rabbits; two of the "swellings" were shown by histopathological examination to be

60 There are organizations opposed to using animals in cancer research. A personal note: I always told my university students that I would far rather have a hundred (or a thousand or a million—or whatever number) rats die than have a child of mine perish from exposure to a chemical of previously unknown cancer potential. This view applies equally for my grandchildren and to their children—and to all people.

61 Tokyo Imperial University is now known as The University of Tokyo.

62 Yamagiwa, K., K. Ichikawa. 1918. Experimental study of the pathogenesis of carcinoma. *Journal of Cancer Research* 3:1–29. The 1918 reference is in English; the original report was made at the Tokyo Medical Society in 1915 and subsequently published in German the same year: Yamagiwa, K., K. Ichikawa. 1915. Expermentelle Studie uber die Pathogenese der Epithelialgeschwülste. *Mitteilungen aus der Medizinischen Fakultät zu der Kaiserlichen Universität zu Tokyo* 15:295–344. The paper published in 1915 establishes scientific priority for Yamagiwa's and Ichikawa's notable contribution.

FIGURE 2.8 *Katsusaburo Yamagiwa, image dated about 1910, was instrumental in providing a method for the production of cancer in experimental animals.*
WIKIMEDIA COMMONS FILE: YAMAGIWA.JPG.

actual metastases. The work of Yamigiwa and Ichikawa was extended by Passey and others, who induced cancer in mouse skin by repeated topical application of ether extracts of tars.[63]

The Yamagiwa and Ichikawa study was the first advance in experimental chemical carcinogenesis. The reader is reminded that prior studies of carcinogenesis resulted from occupational data of men who worked in what proved to be hazardous occupations; these were unplanned experiments of nature. Many scientists tried and failed to induce cancer in experimental animals in the early years of the 20th century. Yamagiwa and Ichikawa succeeded in this

63 Haddow, A., G.A.R. Kon. 1947. Chemistry of carcinogenic compounds. *British Medical Bulletin* 4:314–325. Note: The entire issue of the British Medical Bulletin that contains the Haddow and Kon paper is devoted to chemical carcinogenesis. The entire issue is a remarkable review of the "state of the art" of chemical carcinogenesis in the time shortly after the Millers began their studies.

endeavor by planning their experiments carefully, with talent, diligence and persistence, and succeeded where others failed.[64] For his contribution to the field of cancer research and for his many other contributions, Yamagiwa has been described as the Father of Cancer Research in Japan.

Many honors came to Yamagiwa, but the Nobel Prize escaped him. Yamagiwa was nominated for the Nobel four times and came close to receiving the prize in 1926, but a strange quirk of judgment resulted in the Nobel Prize being awarded in 1927 to Fibiger. Johannes Andreas Grib Fibiger (1867–1928) made the erroneous assertion that stomach cancer in rats was caused by a nematode worm carried to the rats by the common cockroach.[65] While Fibiger's work attracted much attention, it was never repeated,[66] and it must go down in history as an error resulting in deep embarrassment to the Nobel selection committee. At least one historian referred to the 1926 Nobel Prize as a "fiasco".[67] In contrast with the work of Fibiger, many people *have* repeated and extended the work of Yamagiwa—so much for the infallibility of committees in general, and in this case a Nobel Prize committee in particular.

Rabbits are large by comparison with mice. The smaller animals take less laboratory space, are more economic and are simple to use, and thus it was helpful to discover that experimental cancer can be induced in mice. This was shown by Tsutsui, who was in the same University of Tokyo laboratory as Yamagiwa.[68]

By the 1930s, chemical procedures had evolved to the point that individual compounds could be identified and characterized from a complex mixture

64 The historical contribution of Yamagiwa and Ichikawa, as well as the failure of others, is recounted in: Henschen F. 1968. Yamagiwa's tar cancer and its historical significance. *Gann* 59:447–511.

65 Fibiger J. 1913. Untersuchung űber eine Nematode (Spiroptera sp.n.) und deren Fähigkeit, papillomatöse und karzinomatöse Geschwulstbildungen im Magen der Ratte hervorzurufen. *Zeitschrift für Krebsforschung* 13:217–280.

66 Not repeated but an alternative explanation is found in: Hitchcock, C.R., E.T. Bell. 1952. Studies on the nematode, parasite, Gongylonema neoplasticism (Spiroptera neoplasticism) and avitaminosis A in the forestomach of rats: comparison with Fibiger's results. *Journal of the National Cancer Institute* 12:1345–1387. Hitchcock and Bell were both University of Minnesota Medical School faculty and 29% of the infected Norwegian brown rats of their study were obtained in suburban Minneapolis—perhaps an inadvertent commentary on the state of Public Health at the time.

67 Bartholomew, J.R. 1998. Japanese Nobel candidates in the first half of the twentieth century. *Osiris* 13:238–284. This is an outstanding review of the work of Yamagiwa, see pages 253–262.

68 Tsutsui, H. 1918–1919. Über das künstlich erzeugte Cancroid bei der Maus. *Gann* 12:17–21.

of soot and tars. It was not long before polycyclic aromatic hydrocarbons,[69] identified from coal tar and applied to mice, were shown to have the competence to cause cancer. This very important time in the study of cancer-causing chemicals has been reviewed by Ernest Laurence Kennaway (1881–1958)[70] and Alexander Haddow.[71] The polycyclic hydrocarbon 1,2,5,6-dibenzanthracene (also known as dibenz [*a,h*] anthracene) was the first *pure* chemical tested for, and shown to have, the competence to cause cancer.[72] Dibenz [*a,h*] anthracene causes lung, mammary and a number of other cancers in mice. It has never been produced commercially, but it occurs "naturally" in automobile exhausts, tobacco smoke and charcoal-broiled steaks, as well as in coal tar, shale oils, and soots. Starting with two tons of "medium soft pitch" ("pitch" = coal tar), benzo(a)pyrene was the first carcinogen isolated from a complex mixture and shown to be a potent carcinogen.[73] Benzo(a)pyrene is an organic chemical closely similar to dibenz[a,h] anthracene.

A chemical of much importance to Elizabeth and Jim Miller was the derivative of *o*-aminoazotoluene (4-dimethylaminoazobenzene, DAB) also known as butter yellow. It was shown to be carcinogenic by Kinosita.[74] The door was open for the identification of chemical carcinogens.

---

69 Cook, J.W., I. Hieger, E.L. Kennaway, W.V. Mayneord. 1932. The production of cancer by pure hydrocarbons. *Proceedings of the Royal Society of London*, Series B 111:455–484.

70 Kennaway, E.L. 1955. The identification of a carcinogenic compound in coal-tar. *British Medical Journal* 2(4942):749–752.

71 Haddow, A. 1974. Sir Ernest Laurence Kennaway FRS, 1881–1958: Chemical causation of cancer then and today. *Perspectives in Biology and Medicine* 17:543–588.

72 There is a consensus that dibenz [a,h] anthracene has the distinction of being the first pure chemical compound shown to cause cancer, see: Miller, E.C., J.A. Miller. 1979. Milestones in chemical carcinogenesis. *Seminars in Oncology* 6:445–460. Miller, E.C. 1978. Some current perspectives on chemical carcinogenesis in humans and experimental animals: presidential address. *Cancer Research* 38:1479–1496. Miller, J.A. 1994. Brief history of chemical carcinogenesis. *Cancer Letters* 83:9–14. Coombs, M.M. 1980. Chemical carcinogens: a view at the end of the first half century. *Journal of Pathology* 130:117–146. Peacock, P.R. 1944. Fluorescence as an aid to physiology. *Nature* 153:136. Loechler, E.L., B. Henry, K.-Y. Seo. 2002. Cellular responses to chemical carcinogens. In: W.B. Coleman, G.J. Tsongalis (eds) *The Molecular Basis of Human Cancer.* Totowa, New Jersey: Humana Press, pp. 203–222.

73 Cook, J.W., C.L. Hewett, I. Hieger. 1933. The isolation of a cancer-producing hydrocarbon from coal tar. Parts I, II, III. *Journal of the Chemical Society* 395–405.

74 Kinosita R. 1936. Researches on the carcinogenesis of the various chemical substances. *Gann* 30:423–426. (in Japanese). See also, in English, Kinosita, R. 1940. Studies on the cancerogenic azo and related compounds. Yale Journal of Biology and Medicine 12:287–300.: Yoshida, T. 1933. Über die serienweise Vergolgung der Veränderungen der Leber der

The list of pure chemicals that cause cancer grew rapidly by mid-20th century. That list included 2-acetylaminofluorene, ethyl carbamate (urethane), chloroform, carbon tetrachloride, thiourea, thioacetamide, the chemotherapeutic nitrogen mustards, beryllium derivatives, dimethylnitrosamine, cycasin, safrole, aflatoxin B1, mitomycin C and many others.[75]

The World Health Organization, International Agency for Research on Cancer, lists of carcinogens can be accessed at http://monographs.iarc,fr/ENG/classification. The chemicals are listed in groups, *viz.*, Group 1 Carcinogenic to humans, Group 2A Probably carcinogenic to humans, Group 2B Possibly carcinogenic to humans, Group 3 Not classifiable as to carcinogenicity to humans, Group 4 Probably not carcinogenic to humans. The website also lists specific cancers with chemicals that possibly are etiologic to those cancers. Another superb source of information on chemicals and cancer is that of the 13th Report on Carcinogens 2014, published by the National Toxicology Program, Department of Health Education and Welfare, National Institute of Health (USA), Appendix 4. The 13th Report on Carcinogens contains two lists: one is of substances known to be carcinogenic and the other is a list of substances that may reasonably be anticipated to be human carcinogens. The "substances" are of many kinds including metals, pesticides, drugs and natural and synthetic chemicals. I, as the author of this book, wish to add a note of reason to this discussion. Simple exposure to a given listed substance does not guarantee that the individual will succumb to cancer. Readers of this book will recognize that cancer is far more complicated. Factors which contribute to vulnerability include amount of exposure, duration of exposure and individual susceptibility. Even so, a wise person will exercise extreme caution to avoid exposure.

Chemical Structure of Carcinogens

Structural formulae for chemical carcinogens clearly illustrate that they are indeed a diverse group which often share no common structural features.[76] Also, they are diverse in origin: that is, they occur in nature, they result from

experimentellen Hepatomerzeugung durch o-aminoazotoluol. *Transactions of the Japan Pathological Society* 23:636–638.

75 Miller, J.A. 1994. Brief history of chemical carcinogenesis. *Cancer Letters* 83:9–14.

76 Miller, J.A. 1970. Carcinogenesis by chemicals: An overview—G.H.A. Clowes Memorial Lecture. *Cancer Research* 30:559–576. Miller, E.C., J.A. Miller. 1979. Milestones in chemical carcinogenesis. *Seminars in Oncology* 6:445–460.

human activity, they are produced by our gifted synthetic chemists, and some occur because of human (and animal) metabolic activity.

Chemical carcinogens, belonging as they do to more than a dozen different chemical groups, suggest to some that there are as many different carcinogenic processes as there are chemical carcinogens. Were this assumption to be true, the biology of chemical carcinogenesis would be bewilderingly complex, which of course it is, but it is certainly *less* bewildering if one does not have to study potentially hundreds of mechanisms. The research of the Millers, however, led to the realization that, while carcinogens differ, the *manner* in which they cause cancer results from the same biochemical processes. Despite their chemical dissimilarities, the Millers found that *most chemical carcinogens are metabolized to become potent electrophilic*[77] *reactants*. The activated electrophilic chemicals combine chemically with nucleophilic[78] sites on DNA, RNA and protein. Perhaps counter intuitively, few of the diverse chemicals which cause cancer have the intrinsic competence to cause cancer; i.e., they are not carcinogens as such. Chemical carcinogens should more properly be referred to as procarcinogens because they must be metabolically altered into reactive carcinogenic compounds by the host. The reactive chemical then covalently binds with cell macromolecules (DNA, RNA, proteins). This *conversion of non-reactive procarcinogens, by enzymes of the cell, to cancer-causing chemicals (sometimes known as "metabolic activation")* is the area to which the Millers contributed and for which they are best known. As has been said before, and will be said yet many more times, cancer is not a simple problem. The Miller's work focused on how chemicals could become reactive with large molecules of the cell to cause cancer. The resulting change in DNA (or RNA or proteins), while a first critical event, is far from the complete narrative. Regardless of how complex the many steps are that follow leading to clinical cancer, there has to be an original precipitating event and it is there that Elizabeth and Jim Miller contributed so mightily.

The preceding discussion is not to say that direct-acting carcinogens do not exist. To the contrary, a number of carcinogens exist in their reactive forms. This book, however, is about the research peculiar to the Millers and little more will be written here about direct-acting chemical carcinogens.

---

77 Electrophilic: a chemical reactant having an affinity for electrons; pertains to electron acceptance in covalent binding.

78 Nucleophilic: an electron donor in covalent binding.

Carcinogenesis is a Multistep Process

Even in the early days of carcinogenesis studies, it was recognized that complexities existed. A principal finding was that chemicals that caused cancer did not act alone; there was enhancement to the process by other treatments, and carcinogenesis was a multistep process. An example: creosote alone does not cause mouse skin cancer; benzo(a)pyrene does cause skin cancer; benzo(a) pyrene *plus* creosote yields an increase of cancer—the enhancing effect of creosote resulted in it being named a cocarcinogen.[79] Isaac Berenblum showed that the oil derived from seeds of *Croton tiglium*, a leafy shrub or tree found in southeast India, contain a blister-causing substance termed "croton oil". Croton oil is toxic to bacteria, insects, amphibia, fish and mammals; it has been used to induce drastic diarrhea or as a counterirritant although the oil is no longer considered safe for human consumption.[80] Croton oil significantly increased the number of tumors in mouse skin treated with benzo(a)pyrene. The action of croton oil was referred to as cocarcinogenesis.[81] Mottram a few years later showed that benzo(a)pyrene, with as little as a single application to mouse skin, when followed by treatment with croton oil, resulted in skin papillomas thus confirming the observations of Berenblum.[82] The principal promoting substance in the cocarcinogen croton oil was shown to be 12-0-tetradecanoylphorbol-13-acetate (TPA).[83]

---

79 Shear, M.J. 1938. Studies in carcinogenesis. V. Methyl derivatives of 1,2-benzanthracene. *American Journal of Cancer* 33:499–537 (see discussion page 531 concerning an enhanced tumor effect of creosote oil).

80 Some torpedoes were powered by alcohol during World War II. The U.S. Navy added small amounts of croton oil to the alcohol to deter sailors from drinking the torpedo fuel. Enterprising sailors devised small stills to separate the alcohol from the offending (to the sailors) croton oil.

81 Berenblum, I. 1941. The cocarcinogenic action of croton resin. *Cancer Research* 1:44–48. Berenblum, I. 1941. The mechanism of carcinogenesis: A study of the significance of cocarcinogenic action and related phenomena. *Cancer Research* 1:807–814 (reprinted in: CA, *A Cancer Journal for Clinicians* 31:241–253).

82 Mottram, J.C. 1944. A developing factor in experimental blastogenesis. *Journal of Pathology and Bacteriology* 56:181–187.

83 Hecker, E. 1971. Isolation and characterization of the cocarcinogenic principles from croton oil. *Methods in Cancer Research* 6:439–484.

Carcinogenesis is thus a multistage phenomenon, as suggested by the terms initiation and promotion (progression was added later).[84] The exposure of normal cells to a single low dose of a carcinogen results in their conversion to latent tumor cells. This is known as initiation and is probably irreversible because it involves changes in DNA.[85] The target cells are those with the competence to replicate. Subsequent exposure of the latent ("incipient") tumor cells with a promoter causes formation of many tumor nodules that are probably benign. The conversion of latent tumor cells to multiple tumors is known as promotion, and it is believed to be reversible; i.e., removal of the promoter results in the disappearance of the nodules. Promotion involves the stimulation or enhanced expression of mutated genes, i.e., transcriptional[86] activity of DNA sequences that have been altered by binding with metabolically activated carcinogens. The next step and final step in cancer induction is known as progression, and this causes tumors to become more aggressive; i.e., to become more malignant. Progression is measured in a multiplicity of factors involving chromosomal abnormalities, leading to alterations in growth rate, invasiveness, metastatic competence, hormone responsiveness, escape from immune surveillance, changes in cell adhesion molecules, defective apoptosis (i.e., loss of programmed cell death) and loss of a terminally differentiated state.[87] Carcinogenesis in several steps has been demonstrated in multiple test systems, e.g., urinary bladder,[88] mammary cancer[89] and thyroid.[90] This

84 Friedewald, W.F., P. Rous. 1944. The initiating and promoting elements in tumor production. An analysis of the effects of tar, benzpyrene, and methylcholanthrene on rabbit skin. *Journal of Experimental Medicine* 80:101–126.

85 Pitot, H.C. 1993. The molecular biology of carcinogenesis. *Cancer* 72:962–970.

86 Transcription is the process of making a RNA copy of genetic information (DNA). The transcribed RNA may be subsequently translated into a protein gene product.

87 Foulds, L. 1965. Multiple etiologic factors in neoplastic development. *Cancer Research* 25:1330–1347. Sugimura,T., 1992. Multistep carcinogenesis: a 1992 perspective. *Science* 258: 603–607. Nowell, P.C. 2002. Tumor progression: a brief historic perspective. *Seminars in Cancer Biology* 12:261–266. See also: Initiation, latency and progression discussed in Pierce, G.B., I. Damjanov. The pathology of cancer. In: McKinnell, R.G., R.E. Parchment, A.O. Perantoni, I. Damjanov, G.B. Pierce. 2006. *The Biological Basis of Cancer*, 2nd ed., Cambridge and New York, Cambridge University Press, see especially pp. 36–38.

88 Cohen, S.M., M. Arai, J.B. Jacobs, G.H. Friedell. 1979. Promoting effect of saccharin and DL-tryptophan in urinary bladder carcinogenesis. *Cancer Research* 39:1207–1217.

89 Armuth, V., I. Bereblum. 1974. Promotion of mammary carcinogenesis and leukemogenic action by phorbol in virgin female Wistar rats. *Cancer Research* 34:2704–2707.

90 Hall, W.H., F. Bielschowsky. 1949. The development of malignancy in experimentally induced adenomata of the thyroid. *British Journal of Cancer* 3:534–541.

short paragraph should indicate that carcinogenesis is indeed a complex phenomenon—complex but hopefully not bewildering.

Concluding Remarks

An introduction to what a cancer cell is, combined with an awareness of which chemicals cause cancer, does not reveal *how* chemicals cause cancer. An association of exposure to a particular chemical helps identify which chemicals cause cancer. A rhetorical question follows: what does "association" mean in this context? An example: there is an association between benzene and leukemia reported in workers exposed to that chemical.[91] Nothing in that study of association suggests *how* benzene brings about its malignant outcome. With advances in cell biology however, steps were formulated for the transformation of benzene to a cancer-causing substance. Prior to the expression of its carcinogenic potential, benzene must be metabolically changed by a cytochrome P450 (CYP)-dependent oxidation of benzene to benzene oxide.[92] That is the beginning of the "how" of chemical carcinogenesis. The meaning of the "how" will become abundantly clear in the following chapter.

As late as 1953, the Millers stated "it must be admitted that as yet little is known about the fundamental nature of the carcinogenic process, and no tests are available which enable one to identify with any precision successive states in the genesis of tumors. The only certain indicator that the carcinogenic process has occurred or has been in progress in a given tissue is the eventual emergence of a gross tumor. Consequently the task of defining the carcinogenic process is still largely restricted to the correlation of events in the target tissue with the ultimate appearance of gross tumors".[93] Of course the Millers were not alone in puzzling about the cause(s) of cancer. Shortly after World War II, it was noted that "a large number of the carcinogenic compounds *possess no functional group* in their molecules and it is therefore difficult to understand

91 Infante, P.F., R.A. Rinsky, J.K. Wagoner, R.J. Young. 1977. Leukaemia in benzene workers. *Lancet* 2:76–78.

92 Smith, M.T. 2010. Advances in understanding benzene health effects and susceptibility. *Annual Review of Public Health* 31:133–148.

93 Miller, J.A., E.C. Miller. 1953. The carcinogenic aminoazo dyes. *Advances in Cancer Research* 1:339–396.

how they can bring about the malignant change in a cell by a purely chemical interaction".[94] (italics added).

Elizabeth and Jim Miller stated, "The knowledge of the mechanisms of carcinogenesis by chemicals provides a useful basis for approaches to the prevention of human cancer."[95] Elizabeth Cavert Miller and James Alexander Miller were leaders in opening the study and understanding of the intricacies of chemical carcinogenesis. Their contributions to this profoundly important area of cancer research, briefly described here, follow in greater detail in the next chapter.

---

94 Anderson, W. 1947. Physico-chemical aspects of chemical carcinogens. *Nature* 160:892–895.

95 Miller, E.C., J.A. Miller. 1981. Mechanisms of chemical carcinogenesis. Cancer 47:1055–1064.

CHAPTER 3

The Millers and Chemical Carcinogenesis

James A. Miller and Elizabeth C. Miller discovered and developed the important unifying concept that most carcinogenic and mutagenic chemicals are not carcinogenic or mutagenic *per se* but that these compounds must undergo metabolism to reactive electrophilic metabolites that exert their effects by covalently binding to critical sites on cellular macromolecules (DNA, RNA, and protein). James and Elizabeth Miller were the first to point out that "as a general class, chemical carcinogens would appear to be potential mutagens, and the mutagenicity of a chemical carcinogen in a given system will depend on its extent of conversion to electrophilic reactive form(s) and on the access of these active form(s) to the genetic material in the mutagenicity system under study." These discoveries and novel concepts initiated a new era of modern toxicology and were the basis of (1) rapid mutagenicity tests (the Ames test and others) for the screening of potential human carcinogens, (2) research on chemically induced mutations in protooncogenes and tumor suppressor genes, (3) large person-to-person differences in the sensitivity of people to environmental carcinogens (intraindividual differences in the metabolism of carcinogens to their active forms), (4) formation of macromolecular adducts as indicators of cancer risk in molecular epidemiology, and (5) international laws regulating carcinogens in the diet and the environment.

CONNEY, A.H., M.C. POIRIER, Y.-J. SURH, F.F. KADLUBAR. 2009. ELIZABETH CAVERT MILLER (1920–1987) AND JAMES A. MILLER (1915–2000): *A Biographical Memoir. Washington, D.C.:* National Academy of Sciences.

Carcinogens and Metabolism

What kind of research did the Millers do? The epigraph above is a synopsis of the lifetime investigations of Elizabeth C. and James A. Miller. That synopsis is from an absolutely sterling source—The National Academy of Sciences of the United States. Thus, this chapter is unnecessary if the words of the National Academy are read carefully. However, most people would want more than an extremely brief summary of two lifetimes of research and this chapter is written for "most people".

 | DOI 10.1163/9789004286801_004

This chapter begins with the carcinogenesis studies of the Millers. It seems to me that mutation of proto-oncogenes and tumor suppressor genes (epigraph, number 2) is entirely within the main thesis of the Miller laboratory—thus, proto-oncogenes and tumor suppressor genes are also considered in this chapter. The following chapter discusses the Millers' remarkably serendipitous findings in the related areas that evolved from their work (epigraph, numbers 1 and 3 through 5).

Elizabeth and James were superbly trained in biochemistry as undergraduates at the University of Minnesota (Chapter Six) and the University of Pittsburgh (Chapter Seven) respectively, and as graduate students working for their Ph.D. degrees at the University of Wisconsin-Madison. They shared an abiding interest in cancer research in general and a special interest in how chemicals cause cancer. The McArdle Laboratory for Cancer Research (Figure 3.1) and its director from 1946 to 1972, Harold P. Rusch,[1] afforded the Millers "steadfast support and encouragement"[2] that nurtured their best efforts. It was not all that common in those early days for a laboratory and its director to support *both* husband and wife of a collaborating couple[3]—however, Rusch's confidence in the Millers was clearly not misplaced as evidenced by their subsequent collaboration with Rusch and with their major professor, Carl Baumann. The Millers, Rusch and Baumann teamed in studies on the effect of diet on the cancer-causing competence of the aminoazo dyes.[4]

The research of the Millers is sometimes presented in textbooks and reviews without attribution. This lack of attribution to the Millers is certainly not motivated by the wish to minimize their contributions. Rather, failure to cite the work of Elizabeth and James is simply an acknowledgement of the significance and universal acceptance of their scientific contributions—their studies have become an essential doctrine of chemical carcinogenesis. It is not necessary

1 Harold Paul Rusch was born in 1908 and died in 1988. He received his M.D. degree from the University of Wisconsin in 1933. He was the author of *Something Attempted, Something done: A Personal History of Cancer Research at the University of Wisconsin, 1934–1978*. Madison: Wisconsin Medical Alumni Association, 1984.

2 Miller, J.A. 1970. Carcinogenesis by chemicals: An overview—G.H.A. Clowes Memorial Lecture. *Cancer Research* 30:559–576.

3 An Administrative Procedure at the University of Minnesota, effective 2009, provides guidance for conducting a "Spousal/Partner No-search Hire". The presence of University procedures for such hires speaks to the fact that they are no longer rare.

4 Studies subsequently showed that metabolism of a carcinogen may be influenced by a vitamin or other dietary factors, see: Brown, R.R., J.A. Miller, E.C. Miller. 1954. The metabolism of methylated aminoazo dyes. IV. Dietary factors enhancing demethylation in vitro. *Journal of Biological Chemistry* 209:211–222.

FIGURE 3.1 *The McArdle Laboratory for Cancer Research located at the University of Wisconsin-Madison. This is where Elizabeth C. Miller and James A. Miller did their noteworthy cancer research.*

COURTESY OF THE UNIVERSITY OF WISCONSIN-MADISON ARCHIVES, IMAGE #S05538.

to cite Charles Darwin in every instance when considering evolution, nor is it essential to credit Watson and Crick when contemplating the double helix of DNA;[5] in a similar manner, the metabolic activation of carcinogens, the binding of the activated chemical to cellular macromolecules, and the eventual result of cancer have become so axiomatic of modern cancer biology that, some would say, they require no attribution. The author of this book does not subscribe to that notion and believes full credit with references should be supplied when citing this fundamental aspect of cancer biology.

James Miller once lamented that graduate students seem to end up knowing only what has happened in the past five years—or maybe even more recently than that.[6] "Very frequently you will talk to them [i.e., the graduate students] about a certain person's work ten years ago or farther back, and they're blank." I hope that this chapter and this book will help students and others, including cancer research professionals, to escape the "blank" that might occur with the mention of the half-century old or older work of Elizabeth and James Miller. Their studies in the 1940s and the subsequent years provided the enduring and substantial foundation for the contemporary understanding of how chemicals cause cancer.

Although a number of chemicals are known to, or thought to, cause cancer[7] it would be useful to know what *proportion* of cancer is caused by chemicals. Elizabeth Cavert Miller addressed this issue in her 1977 Presidential Address presented to the 66th Annual Meeting of the American Association for Cancer Research[8] in Denver, Colorado. Elizabeth stated that environmental chemicals

5 Evolution was not unheard of prior to Darwin nor did chemical studies of DNA begin with Watson and Crick. Similarly, Elizabeth C. Miller and James A. Miller did not *begin* the study of chemical carcinogenesis—rather their scientific contributions had such insight, were so thorough and had such originality and depth that at times, it would seem that they started an entire discipline in cancer research—it may seem that way but in reality, they were not alone and neither were Darwin, Watson and Crick nor other modern scientific investigators.

6 Smail, L. 1985. Oral History Program Interview with Elizabeth C. Miller, University of Wisconsin—Madison: Archives History Program.

7 No one knows exactly how many chemicals cause cancer. The Thirteenth Report on Carcinogens of the National Toxicology Program dated 2014 contains over 200 entries but this "may constitute only a fraction of actual human carcinogens." This is because the National Toxicology Program only lists substances for which relevant data exist and have been reviewed and found to meet listing criteria." See Introduction, *Report on Carcinogenesis*, Thirteenth Edition, National Toxicology Program (Appendix 4).

8 The American Association for Cancer Research, Inc. is the oldest and largest association of cancer researchers in the world. Its headquarters are in Philadelphia, Pennsylvania.

were thought to be "major factors in the causation of human cancer"[9] because of the then unknown but low impact of viruses and the relatively minor role of ionizing radiation. She could have added that "heritable cancers are uncommon".[10] Thus, if viruses, radiation, and heredity are relatively uncommon causes of cancer, what is left?[11] Chemicals emerged then as being very important causes of cancer. Of course, Elizabeth's remarks were made a number years ago (1977). Certainly some viruses are now known to be important cancer-causing agents. Even school children are aware of radiation-caused cancer among Hiroshima and Nagasaki survivors, Chernobyl citizens and, perhaps more recently, some people who lived very close to the Fukushima Daiichi Nuclear Power Facility. Hereditary cancer, while relatively uncommon, is especially important in childhood malignancies.

Chemically-induced cancer, in the recent past as well as in the 21st Century, is considered by most investigators a major cause of cancer. One needs to look no further than lung cancer, which is obviously chemically induced. In 2012, 29% of cancer deaths in American males and 26% of cancer deaths in American women were attributed to lung cancer caused by chemicals in tobacco smoke.[12]

The Oxford epidemiologists Richard Doll and Richard Peto state that "cancer is largely an avoidable disease" with perhaps some 80% to 90% caused by "environmental factors". Note here that the "environment" consists of not only exposure to chemicals but also to man-made and natural radiation, diet and other factors. Exactly how much of the 80% to 90% is due strictly to chemical causation is unknown.[13,14]

---

9 Elizabeth would surely have welcomed the report of the President's Cancer Panel: Reuben, S.H. 2010. *Reducing Environmental Cancer Risk. What can we do now?* Bethesda, Maryland: US Department of Health and Human Services. Please note footnote 12 of this chapter.

10 Knudson, A.G. 1985. Hereditary cancer, oncogenes, and antioncogenes. *Cancer Research* 45:1437–1443.

11 Bad luck remains and it is receiving contemporary attention (see Chapter Two).

12 Siegel, R., D. Naishadham, A. Jemal. 2012. Cancer statistics, 2012. *CA A Cancer Journal for Clinicians* 62:10–29. See also: Jemal, A., M.J. Thun, L.A.G. Ries, H.L. Howe, H.K. Weir, M.M. Center, E. Ward, X.-C. Wu, C. Eheman, R. Anderson, U.A. Ajani, B. Kohler, B.K. Edwards. 2008. Annual report to the nation on the status of cancer, 1975–2005, featuring trends in lung cancer, tobacco use, and tobacco control. *JNCI* 100:1672–1694.

13 Doll, R., R. Peto. 1981. The causes of cancer: Quantitative estimates of avoidable risks of cancer in the United States today. *Journal of the National Cancer Institute* 66:1191–1308.

14 The previous chapter contains a discussion of environmental factors/life style and cancer; as was described, most of these cancer-causing conditions work through chemical change.

Augmenting the notion that chemicals are an important etiological factor is the remarkably disparate prevalence of cancer among nations. Stomach, breast, colon, esophagus, lung, cervix and prostate cancers differ tremendously in different regions of the world.[15] The epidemiological observations strongly implicate the causality of cancer by environmental factors—and in this context, chemicals are thought to be a major culprit.

The President's Cancer Panel in 2012 stated that "The panel was particularly concerned to find that the true burden of environmentally induced cancer has been *grossly under estimated.* With nearly 80,000 chemicals on the market in the United States, many of which are used by millions of Americans in their daily lives and are *un- or understudied and largely unregulated,* exposure to potential environmental carcinogens is widespread" (italics added).[16] The Cancer Panel noted that 300 chemical contaminants, with as yet unknown biological effect, have been detected in neonatal blood—this is a situation of "pre-polluted" newborn babies. Perhaps related to pre-pollution of our children is increasing cancer incidence for "unknown" reasons. Cancer remains a major concern with much needed increased awareness of environmental chemicals and their control.[17]

Occam's (Ockham) Sharp Razor

Chemical carcinogens are found in a bewildering assortment of chemical structures that seem to have no commonality other than the capacity to cause cancer. The conundrum of great diversity has the added complexity that many compounds that appear to be structurally related or similar, often very similar, to known chemical carcinogens have no cancer-causing potential at all.

---

15 Jemal,, A., E.M. Ward, M.J. Thun. 2005. Cancer Statistics, pp. 226–241, In: V.T. DeVita Jr, S. Hellman, S.A. Rosenberg. (eds.) *Cancer, Principles & Practice of Oncology*, Philadelphia: Lippincott, Williams & Wilkins.

16 Covering letter to the President of the United States, signed by LaSalle, D., Leffall, M.D., Chair and Margaret L. Kripke, PhD, 2010, *Reducing Environmental Cancer Risk*, President's Cancer Panel 2008–2009 Annual Report. The report is available on line at http://pcp.cancer.gov or hard copy may be ordered by writing to: President's Cancer Panel, 6116 Executive Boulevard, Suite 220, Bethesda, MD 20892.

17 No specific percentage of cancer emerging from exposure to chemicals has been cited. This omission is due in part to the linkage of chemical carcinogenesis with genetics and lifestyle. Cancer is complex and it is unlikely that a specific percentage of cancer caused by chemicals will ever be known.

The Millers raised the question "How do such structurally varied compounds cause normal cells to become cancer cells?"[18] to which could be added the related question of why some virtually indistinguishable chemicals fail to have comparable cancer-causing competence. With the enthusiasm of youth and the strength that emerges from a sound educational foundation, the Millers embarked on a lifetime investigating these issues in chemical carcinogenesis.

If cancer were to be caused differently by each chemical carcinogen, then the problem of molecular induction of cancer becomes essentially insurmountable, for there would be as many mechanisms of cancer production as there are chemical carcinogens. The team of Elizabeth and James Miller, as other scientists, was faced with this problem. No one at this point can surmise exactly what they were thinking concerning the near impossibility of coming up with hundreds (perhaps thousands?) of different mechanisms on how carcinogens interact with cells to result in cancer. What is actually known, how ever, is that they adopted a pragmatic approach early in their studies. They phrased their pragmatism in this way: "But, for the time being at least, it seems useful to employ the hypothesis that all these agents subsequently produce essentially the same biochemical changes leading to tumor formation."[19] In other words, if the problem seems too complex, then make the assumption that it is not as complex as it seems. This assumption permits research to focus on one or a few chemicals and hope for the simpler solution, with the anticipation that findings from the focus will be applicable to most or all of the remaining chemicals, this was later proven to be true. This was, de facto, a fortuitously insightful use of Occam's razor.[20]

---

18 Miller, J.A. 1994. Research in chemical carcinogenesis with Elizabeth Miller—a trail of discovery with our associates. *Drug Metabolism Reviews* 26:1–36.

19 Miller, J.A., E.C. Miller. 1953. The carcinogenic aminoazo dyes. In: J.P. Greenstein, A. Haddow (eds.). *Advances in Cancer Research*, Volume 1. New York: Academic Press, pp. 330–396.

20 The Miller's "simple solution" is a variant of Occam's (also spelled Ockham) razor "*pluralitas non est ponenda sine necessitate*", i.e., plurality should not be posited without necessity. The "razor" is named for the 13th century Franciscan philosopher and sometime teacher at Oxford University, William, ca. 1285–1349. William was born in Occam, Surrey, England. Another example that recalls Occam's razor: "It is our belief that the basic knowledge and methodology that will have been gained in reaching these objectives for just one carcinogenic process and for just one tumor will in turn make it far easier to discover the nature of other tumors and of other carcinogenic processes." Miller, J.A., E.C. Miller. 1960. Biochemical concepts of carcinogenesis. *Canadian Cancer Conference* 4:57–79.

Cancer-Causing Dyes

A particularly favorable example on which the Millers could focus was a group of aminoazo dyes. Both Elizabeth and Jim had experience with p-dimethylaminoazobenzene (DAB) while working on their doctoral dissertations.[21] Specific, but not all, aminoazo dyes induce both primary liver cancer and metastatic liver cancer in the lungs and mesentery of rats. Isolation and dissection of liver is uncomplicated,[22] and biochemical analysis of both the liver and the azo dye is relatively simple. Consequently, much of the Millers' early work was centered on the induction of liver cancer by aminoazo dyes.

The hepatocarcinogenicity of azo dyes[23] was first reported in 1933 by two Japanese investigators, Tomizu Yoshida[24] (1903–1974) and his mentor Takaoki Sasaki (1878–1966), who produced malignant liver tumors in rats with prolonged feeding of *o*-aminoazotoluene[25] (AAT).[26] The Yoshida and Sasaki study was the first known instance of a specific chemical causing cancer in an *internal* organ, the liver, an organ distant from the site of application. Kinosita reported that an isomer[27] of AAT, dimethylaminoazobenzene (DAB),[28]

---

21 Miller, J.A. 1994. Research in chemical carcinogenesis with Elizabeth Miller—a trail of discovery with our associates. *Drug Metabolism Reviews* 26:1–36.

22 Compare isolation of an intact liver with dissection of nerve tissue or pancreatic islet cells—this is what is meant by "uncomplicated" dissection.

23 There are at least 3000 kinds of azo dyes which were formerly used in textiles, papermaking, cosmetics, pharmacology and food. Today, azo dyes are used primarly as textile dyes. Under anaerobic conditions, some gut and environmental microorganisms may reduce azo dyes to form aromatic amines which may be toxic, mutagenic and carcinogenic. See: Chung, K.-T. 2000. Mutagenicity and carcinogenicity of aromatic amines produced from azo dyes. *Journal of Environmental Science and Health,* Part C, 18:51–74. Note concerning the term "hepatocarcinogenicity", the prefix *hepato-* is derived from both the Latin and Greek words for liver and in contemporary language refers to liver, thus hepatocarcinogenicity relates to causes of liver cancer.

24 Professor of Pathology, Tokyo University, Japan.

25 For fastidious organic chemists, AAT is also known as 2',3-dimethyl-4-aminoazobenzene.

26 Yoshida, T. 1933. Über die serienweise Verfolgung der Veränderungen der Leber der experimentellen Hepatomerzeugung durch o-Aminoazotoluol. *Transactions of the Japanese Pathological Society* 23:636–638; Sasaki, T., T. Yoshida. 1935. Experimentelle Erzeugung des Leberkarcinoms durch Fütterung mit o-Amidoazotoluol. *Virchows Archiv für Pathologische Anatomie und Physiologie und für Klinische Medizin* 295:175–200.

27 Isomers are chemical compounds that have the same molecular formula but are different compounds due to their chemical structure. AAT and DAB are isomeric.

28 The common name for DAB at the time of these studies was "butter yellow". The dye was used as a food additive until the study of Opie, E.L. 1944. The pathogenesis of tumors of the liver produced by butter yellow. *Journal of Experimental Medicine* 80:231–246.

also was a chemical carcinogen.[29] Curiously, German scientists were unable to confirm the Japanese work despite continued success by the Japanese.[30] Elizabeth's major professor, Carl Baumann, scrutinized the conflicting data and judged that the differences could be attributed to the diet fed to the rats—the Germans provided more nutritious rations to the experimental animals; the Japanese animal diet was nutrient-poor.[31] Baumann's interest in dietary factors and carcinogenesis was probably augmented by discovery of this fact.[32] James Miller's first scientific paper (other than the 1934 ultramicroscope study published while he was in High School, Chapter 7) was on the effect of diet on hepatic tumors in rodents.[33] Miller and his co-authors reported that rats were protected to a significant extent from tumors caused by DAB if their basal diet was augmented by both high quality protein and vitamin B complex, particularly riboflavin. Elizabeth's doctoral dissertation (Chapter 7) similarly grew from Baumann's early interest in diet and cancer.

From the onset, Elizabeth and James were interested in the metabolism[34] of chemicals that cause cancer, i.e., how the chemicals are processed by animal

---

29 Kinosita, R. 1936. Researches on the cancerogenesis of the various chemical substances. *Gann* 30:423–426. See also: Kinosita R. 1937. Special report: Studies on the cancerogenic chemical substances. *Nippon Byori Gakkai Kaishi* (also known as *Transactiones Societatis Pathologicae Japonicae*) 27:665–727. Kinosita was at the Pathological Institute, Osaka Imperial University.

30 Fischer-Washels, B. 1936. Die experimentelle Erzengung des malignen Leberzellenadenoms durch o-Amidoazotoluol. *Centralblatt für Allgemeine Pathologie und Pathologische Anatomie* 66:359–360. Heep, W. 1936. Zellwucherungen in Leber und Schilddrüse durch enterale oder parenterale Zufuhr von o-Amidoazotoluol. *Frankfurter Zeitschrift für Pathologie* 50:48–62.

31 The Japanese reported that DAB was a carcinogen to rats; this was not confirmed by German scientists. Is the issue of carcinogenicity still unclear? Absolutely not. The recent 2005 *Eleventh Report on Carcinogens* published by the National Toxicology Program, U.S. Public Health Service, lists DAB as a carcinogen to mice and rats and can be "*reasonably anticipated to be a human carcinogen*" (italics added).

32 Baumann was not alone in his interest in diet and cancer; so too were the Japanese, see: Nakahara, W., K. Mori, T. Fujiwara. 1939. Inhibition of experimental production of liver cancer by liver feeding: A study in nutrition. *Gann* 33:406–427.

33 Miller, J.A., D.L. Miner, H.P. Rusch, C.A. Baumann. 1941. Diet and hepatic tumor formation. *Cancer Research* 1:699–708.

34 See: Mueller, G.C., J.A. Miller. 1948. The metabolism of 4-dimethylaminoazobenzene by rat liver homogenates. *Journal of Biological Chemistry* 176:535–544. Conney, A.H., J.A. Miller, E.C. Miller. 1956. The metabolism of methylated aminoazo dyes. V. Evidence for the induction of enzyme synthesis in the rat by 3-methylcholanthrene. *Cancer Research* 16:450–459. The Millers were not alone in the study of carcinogen metabolism; see, Heidelberger, C., M.R. Kirk, M.S. Perkins. 1948. The metabolic degradation in the mouse

organs. They already had experience with aminoazo dyes in the laboratory, and it was natural that they would continue to study these chemicals. *The azo dyes do not cause cancer directly*. Feeding the dyes to the rats did not result in cancer of the mouth or digestive tract. Once absorbed, however, the chemicals did cause cancer in the rat's liver. The Millers hypothesized that the difference in cancer-causing competence of these azo dyes must therefore ensue from internal metabolic changes to the chemical by the host.

Accordingly, the Millers decided to study the metabolism of the azo dye to ascertain if one or more of its metabolic forms had the capacity to lead to cancer. During that time, another group at the Memorial Hospital and Sloan-Kettering Institute for Cancer Research in New York City was studying the metabolic fate of DAB. The New York group reported isolation of several *urinary* derivatives of DAB after subcutaneous injection of the chemical.[35] This may have spurred the Millers to study the liver of treated animals for metabolic products of DAB. The livers of rats fed DAB contained that parent compound as well as its *mono* methyl (i.e., demethylated) derivative, p-monomethylaminoazobenzene (MAB), and the non-methylated p-aminoazobenzene (AB). As can be seen, rats' livers were able to stepwise remove methyl groups from DAB. Recall the prefix of the parent compound "dimethyl"; removal of one methyl group yields monomethyl as in monomethyl-4-aminoazobenzene (MAB). The Millers and Baumann wrote that demethylation "constitutes a major pathway in the metabolism of methylated dyes".[36] Methylation, i.e., the addition of a methyl group, also occurred in rat livers. The livers of rats fed the mono-methyl compound were found to contain the dimethyl- form showing that rat liver has the competence to both methylate and demethylate.[37] This was probably a major pathway and a fertile area to study tissue and chemical interaction.

In 1947, Elizabeth and James Miller suggested new terms relating to carcinogen metabolism.[38] They hypothesized that a compound (or compounds)

---

of dibenzanthracene labeled in the 9 and 10 positions with carbon 14. *Cancer* 1:261–275. Weisburger, E.K., J.H. Weisburger. 1958. Chemistry, carcinogenicity and metabolism of 2-fluorenamine and related compounds. *Advances in Cancer Research* 5:331–431.

35 Stephenson, E.S., K. Dobriner, C.P. Rhoads. 1942. The metabolism of dimethylaminoazobenzene (butter yellow) in rats. *Cancer Research* 2:160–167.

36 Miller, J.A., E.C. Miller, C.A. Baumann. 1945. On the methylation and demethylation of certain carcinogenic azo dyes in the rat. *Cancer Research* 5:162–168.

37 Miller, J.A. 1994. Brief history of chemical carcinogenesis. *Cancer Letters* 83:9–14.

38 Miller, E.C., J.A. Miller. 1947. The presence and significance of bound aminoazo dyes in the livers of rats fed p-dimethylaminoazobenzene. *Cancer Research* 7:468–480. See also: Miller, E.C, J.A. Miller. 1952. In Vivo combinations between carcinogens and tissue constituents and their possible role in carcinogenesis. *Cancer Research* 12:547–556.

derived from an azo dye was directly responsible for the initiation of chemical carcinogenesis. As such, one would consider that the azo dye carcinogen in the diet, referred to by the Millers as the *parent* compound, was metabolically converted in the body to a *derivative* that was intimately involved in the malignant process. The derived chemical was referred to as a *primary carcinogen* and it was this chemical that interacted with a cell to cause cancer.

A working hypothesis of the 1947 paper was that the primary azo dye carcinogen is p-dimethylaminoazobenzene (DAB) or its metabolite, p-monomethylaminoazobenzene (MAB) or a mixture of the dyes or a compound closely related to the dyes. Their researches led directly to the finding that rats fed DAB have liver cancer with aminoazo dye derivatives (metabolites) *bound tightly* to a cell constituent, liver protein. The bound dyes were p-monomethylaminoazobenzene and p-aminoazobenzene, as well as an unidentified dye. Covalent binding[39] of these two metabolic derivatives to rat liver proteins was reported—no free (i.e., unbound) dye was detected. The notion of "co-valent" binding followed from the observation that the bound dye could not be freed from the protein by any means other than chemical destruction of the protein. The early Miller investigations provided strong support, and indeed the first evidence, for the notion that cancer results from the metabolic activation of a potential carcinogen followed by the interaction (covalent binding)[40] of the metabolic derivative with large informational macromolecules.[41] The Millers stated: "This appears to be the first report of a firm combination *in vivo* of derivatives of a carcinogen with a cellular constituent of high molecular weight." Jim Miller remembered many years later that he and Elizabeth were astonished by the finding of a carcinogen derivative chemically bonded with cell protein!

The contemporary reader might acknowledge astonishment with protein binding, but a far more penetrating question would be whether covalent binding with DNA occurs. The Millers did then consider activated carcinogen binding with other cellular constituents, including DNA. The timing of the 1947

39 Binding of activated chemical carcinogens occurs in other body cells. These cell/carcinogen complexes become biomarkers for exposure to specific agents. Detection and quantitation of biomarkers is known as molecular epidemiology, discussed in the next chapter.

40 A covalent bond is formed by the sharing of two electrons by two atoms.

41 Informational macromolecules are protein, RNA and DNA. Initially, the Millers reported that the activated chemical bound with protein (which it did). Later, they extended their observations to include both RNA and DNA.

paper was *prior* to the 1953 publication of the Watson and Crick[42] seminal paper that resulted in the nearly ubiquitous contemporary focus on DNA as it relates to genetic stability or genetic change. Elizabeth and James sought to detect covalent binding of an activated azo dye to liver DNA, and accordingly, they made a preparation of total nucleic acid from their azo dye-treated hepatic tissue. Presence of the dye was thought to be to detectable by a pink color (remember that the azo compounds are dyes). Since no pink color was found, the Millers concluded that no azo dye was bound to the DNA. Now many years later, with hindsight, I lament: "what a pity". Jim acknowledged that "As time quickly proved the conclusion from this experiment was erroneous."[43] The error resulted because the DNA-dye complex did not produce the expected or hoped for pink color.[44] Allan H. Conney, in a personal communication of March 5, 2013, described the failure to obtain a pink color as "bad luck" and stated that if the color had been obtained, the Millers would have received the Nobel Prize.

However, the Millers wrote (in 1952, this too was *prior* to the Watson and Crick paper): "Since protein preparations such as these are a mixture of proteins and nucleoproteins, the dye might be combined with either the protein or nucleic acid moieties or both." Further, they stated "*reaction with DNA is a particularly attractive model for carcinogenesis*" (italics added).[45] Later (1966), the Millers wrote "Alterations in DNA can affect the potential metabolic properties of the cell and permit the transmission of altered genomes to subsequent cell generations. Alterations in RNA or protein can also clearly produce profound effects within a particular cell, but the means of *propagation* (italics added) of these effects to progeny cells is less clear." Elizabeth, in her 1977 presidential

42 Watson, J.D., F. Crick. 1953. Molecular structure of nucleic acids: A structure for deoxyribose nucleic acid. *Nature* 171:737–738.

43 Miller, J.A. 1998. The metabolism of xenobiotics to reactive electrophiles in chemical carcinogenesis and mutagenesis: A collaboration with Elizabeth Cavert Miller and our associates. *Drug Metabolism Reviews* 30:645–674. Concerning the binding of DNA with activated chemical carcinogens, see: Warwick, G.P., J.J. Roberts. 1966. The covalent binding of metabolites of dimethylaminoazobenzene, ß-naphthylamine and aniline to nucleic acids in *vivo*. *International Journal of Cancer* 1:179–196. Warwick, G.P, J.J. Roberts. 1967. Persistent binding of butter yellow metabolites to rat liver DNA. *Nature* 21:1206–1207.

44 Lin, J.-K., J.A. Miller, E.C. Miller. 1975. Structures of hepatic nucleic acid-bound dyes in rats given the carcinogen *N*-methyl-4-aminoazobenzene. *Cancer Research* 35:844–850.

45 Miller, E.C., J.A. Miller. 1952. *In vivo* combinations between carcinogens and tissue constituents and their possible role in carcinogenesis. *Cancer Research* 12:547–556. Also see: Miller, E.C., J.A. Miller. 1966. Mechanisms of chemical carcinogenesis: Nature of proximate carcinogens and interactions with macromolecules. *Pharmacological Reviews* 18:805–838.

address to the American Association for Cancer Research in Denver, Colorado, stated "The recognition of the central roles of DNA's [sic] as storehouses of genetic information and of RNA's [sic] in the translation of the genetic information for the synthesis of cellular proteins provided new perspectives on the possible roles of nucleic acid-carcinogen adducts in carcinogenesis and provided an impetus to search for such derivatives."

There is no question that the Millers' initial observation of activated carcinogen binding to protein is correct, but later (as shown in the previous paragraph) they clearly stated that carcinogen binding to DNA would probably be more important in the genesis of cancer than binding with other large molecules. In my opinion, the Millers' early work on covalent binding of activated carcinogens to macromolecules would probably be better known (at least in the years immediately following the *Nature* paper of Watson and Crick) if they had detected covalent binding to DNA. It was not their intellect that failed—rather, it was the limitation of their analytical chemical procedure that was wanting.

The Millers noted that no bound dye was found in any tissue that was unaffected by the azo dye carcinogenesis. These tumor-free tissues lacking azo-bound dye included small intestine, kidney, spleen, lung, heart and skeletal muscle. The Millers stated, "The virtual absence of the bound dyes from all of the rat tissues studied except the liver is an obvious correlation with the high specificity of the carcinogenic aminoazo dyes in producing only liver tumors."

The Millers also reported that no bound dye was found in the livers of species which are *resistant* to azo dye carcinogenesis; these included livers of guinea pigs, rabbits, cotton rats,[46] and chickens. The Millers recognized the importance of their discoveries and so stated: "The apparent correlations observed between the level of bound dye in a liver and the likelihood of a tumor developing in that liver under the varied conditions studied suggest that the bound dyes have an important place in the carcinogenic process induced by p-dimethylaminoazobenzene."

Recall that it was thought that chemical carcinogenesis might result from one or a few mechanisms or it could (perish the thought) ensue from as many mechanisms as there are chemical carcinogens. The historic first findings with amino azo dyes and their metabolic derivatives covalently bound to macromolecules were followed by studies of macromolecular-bound carcinogenic derivatives of other chemicals. Occam's razor does indeed work, at least some of the time.

---

46 A cotton rat, so called because it builds a nest of cotton, is any member of the genus *Sigmodon*.

Elizabeth shaved a patch of skin on the back of mice and applied benzo(a) pyrene for five days. On the sixth day, she harvested the treated skin and isolated skin proteins. She reported protein-bound derivatives of benzo(a)pyrene—the binding of the metabolite with the skin protein withstood multiple washings thus establishing that the stable binding was apparently covalent. Biologists, in general, do not look for unique phenomena. Rather, most seek new information that can be applied to many or most organisms. Such generalizations are of fundamental importance to biology and clearly transcend the less important unique observations. It is easier to learn laws and how they operate than to scrutinize endless lists of discrete mechanisms. The Millers were composing biological law that would likely stand for all time. Elizabeth Cavert Miller added benzo(a)pyrene to dimethylaminoazobenzene as compounds that were metabolically changed to permit covalent binding to cellular macromolecules. This research of the Millers, on two distinct classes of chemical carcinogens, provided the *first evidence* that covalent binding of a carcinogen metabolite to a host macromolecule is an early and essential step in chemical carcinogenesis. Other aminoazo dyes, mycotoxins, ethionine, urethane, many allylarenes,[47] several polycyclic aromatic hydrocarbons, 2-acetylaminofluorene (2-AAF, discussed in the next section), etc. were found to act similarly.[48] The generality of the notion that carcinogens must be metabolically changed prior to covalent binding with cell macromolecules was moving from a reasonable hypothesis to a well-established principle.

I hope the reader will permit a personal comment on the 1952 Miller paper.[49] I have read and reviewed many scientific studies for publication. I have found few indeed that are as data-dense as this paper. The data from the 1952 Millers'

---

47 Allylarenes are benzene ring-containing (aromatic) compounds linked with the monovalent allyl group.

48 Miller, E.C. 1951. Studies on the formation of protein-bound derivatives of 3,4-benzpyrene in the epidermal fraction of mouse skin. *Cancer Research* 11:100–108. Miller, E.C., J.A. Miller. 1952. In vivo combinations between carcinogens and tissue constituents and their possible role in carcinogenesis. *Cancer Research* 12:547–556. Miller, E.C., J.A. Miller. 1966. Mechanisms of chemical carcinogenesis: Nature of proximate carcinogens and interactions with macromolecules. *Pharmacological Reviews* 18:805–836. See also studies of fluorenamine metabolism at the National Cancer Institute: Weisburger, E.K., J.H. Weisburger. 1958. Carcinogenicity and metabolism of 2-fluorenamine and related compounds. *Advances in Cancer Research* 5:331–431; Heidelberger, C. 1964. Studies on the molecular mechanisms of hydrocarbon carcinogenesis. *Journal of Cellular and Comparative Physiology*, 64 (Supplement 1):129–148.

49 Miller, E.C., J.A. Miller. 1952. In vivo combinations between carcinogens and tissue constituents and their possible role in carcinogenesis. *Cancer Research* 12:547–556.

paper form a compelling argument for the thesis that derivatives of carcinogens can be detected tightly bound to a cellular component—in this case, hepatic protein. Elizabeth, in a later paper, unpretentiously stated that she thought that the binding reaction of the carcinogen metabolite to a cellular protein "might play an important part in the induction of tumors." It is appropriate to use "unpretentious" in describing Elizabeth's modesty in this quotation. Elizabeth was a Finney-Howell post-doctoral fellow when the paper was published and she was but 27 years of age. That 1952 paper of the Millers which stated that "alterations in DNA" permit the transmission of changed genomes to subsequent cell generations is, in my opinion, a noteworthy herald to the now famous 1953 paper of Watson and Crick, which elucidated the molecular structure of DNA.[50]

Metabolism of a carcinogen with covalent binding of its activated metabolite led to the discovery that this binding is the critical reaction by which most carcinogens induce cancer. This unifying concept of chemical carcinogenesis may be the most lasting of the scientific contributions of Elizabeth and James Miller.[51]

Can Chemical Structure Predict Carcinogenicity?

James Miller wanted to investigate a series of azo dyes that varied structurally from each other in subtle ways. Accordingly, he sought financial support from Carl Baumann, who provided funds to buy the necessary chemicals. James proceeded to make the variant azo dyes. As James modestly stated in an interview years later, "chemically it's pretty simple to do—to alter the structure" of azo dyes. After synthesis and purification, he fed the chemically altered azo dyes to rats. This was no short-term survey. The Millers and Baumann produced 29 closely related compounds (closely related by chemical structure) which were

50 Watson, J.D., F.H.C. Crick. 1953. Molecular structure of nucleic acids. A structure for deoxyribose nucleic acid. *Nature* 171:737–738.

51 For an excellent early review of these studies, see: Miller, J.A., E.C. Miller. 1953. The carcinogenic aminoazo dyes. *Advances in Cancer Research* 1:339–396. Concerning the discussion of DNA above, I cite: "The bound dyes do not appear to be bound to the nucleic acids present in the liver protein preparation… the finding that the dyes are not bound to the major share of the nucleic acids does not exclude them from being bound to the protein moieties of nucleoproteins."

derivatives of DAB.[52,53] The studies took months to complete. One of the compounds, m'-methyl-p-dimethylaminoazobenzene, was shown to have more cancer-causing potential than the parent compound; in fact, it was "proved to be the most potent carcinogenic azo dye hitherto reported for the liver of the rat."[54] One hundred percent of the rats fed the m'-methyl derivative for 2 ½ months developed liver cancer. Of the other compounds, only p-monomethylaminoazobenzene (MAB) showed significant cancer-causing competence. The remaining derived azo dyes, although structurally similar to DAB, had less or even no carcinogenic effect.

As the author of this book, I wish to offer a personal comment. I am not amazed that wildly variant chemical substances can cause cancer. I suppose that I generalize this since certain viruses, specific kinds of radiation and some chemicals are so very dissimilar but all can, under appropriate circumstances, cause cancer; thus, I am not then amazed that *dissimilar* chemicals can also cause cancer. However, even though I have known of the Millers' work for nearly a half century, I am still amazed that structurally very similar chemical compounds, compounds that differ from each other in subtle and almost indistinguishable ways, can have such variant potential to cause cancer. I would have been sorely tempted at the time of the Millers' work to throw up my hands and walk out of the laboratory. How grateful humankind should be, as I am, that Elizabeth and James did *not* walk away.

Metabolic Studies of Other Chemicals

2-acetylaminofluorene (AAF) is a potent carcinogen to the liver, bladder and many other tissues in the rat, mouse, hamster, rabbit, fowl, cat and dog, but not the guinea pig nor the cotton rat. Vulnerability to AAF varies by animal strain and with diet and hormones. Tumors induced by AAF appear at sites distant from the route of administration. Metabolic and carcinogenicity investigations of AAF have been studied extensively and reviewed by Jon Hans Weisburger and Elizabeth K. Weisburger.[55]

---

52 Miller, J.A., C.A. Baumann. 1945. The carcinogenicity of certain azo dyes related to p-dimethylaminoazobenzene. *Cancer Research* 5:227–234.

53 Miller, J.A., E.C. Miller. 1948. The carcinogenicity of certain derivatives of p-dimethylaminoazobenzene in the rat. *Journal of Experimental Medicine* 87:139–159.

54 Giese, J., J.A. Miller, C.A. Baumann. 1945. The carcinogenicity of m'-methyl-p-dimethylaminoazobenzene. *Cancer Research* 5:337–340.

55 Weisburger, E.K., Weisburger, J.H. 1958. Chemistry, carcinogenicity and metabolism of 2-fluroenamine and related compounds. *Advances in Cancer Research* 5:331–431.

The first step in the metabolic activation of AAF is N-hydroxylation by an enzyme of the cytochrome P450 system (discussed later in this chapter) of the endoplasmic reticulum[56] that results in N-hydroxy-2-acetylaminofluorene (N-hydroxy-AAF) as shown by John Cramer[57] of the Miller laboratory. The N-hydroxy form is a stronger carcinogen than the parent compound, as evidenced by its action in a wider range of tissues and species. Further metabolism of the N-hydroxy-AAF resulted in compounds of yet greater toxicity. The Millers and Cramer were thus the first to demonstrate that metabolism of a carcinogen could result in a compound with increased carcinogenicity, i.e., the metabolized carcinogen had a greater cancer-causing potential than the parent carcinogen.

Furthermore, the Millers added to their nomenclature the terms *proximate* and *ultimate* carcinogen. As would be predicted with the benefit of more than a half century of hindsight, both AAF and the N-hydroxy metabolite of AAF give rise to derivatives that bind to protein, RNA and DNA in the living animal. However, because no such derivatives were formed in laboratory glassware (*in vitro*), the Millers postulated that additional metabolism was required prior to the formation of the ultimate carcinogen(s) of AAF. This indeed proved to be the case with their discovery of multiple electrophilic[58] metabolites, including 2 acetyl-aminofluorene- N-sulfate. The AAF-N-sulfate compound has a high mutagenic potential[59] in a DNA-transforming system,[60] and is probably the

56 The endoplasmic reticulum, found in the non-nuclear part of cells of all higher plants and animals, is a complex of membranes. Some of the membranes contain granules and other membranes are smooth (granule-free). Only the electron microscope permits visualization of the reticulum structure. Synthesis of various substances takes place in the endoplasmic reticulum.

57 John Wesley Cramer was born in 1928, received his PhD in biochemistry from the University of Wisconsin, Madison in 1955, and was a professor of pharmacology, School of Medicine, Mercer University, Macon, Georgia. Cramer et al. was the first unequivocal demonstration of N-hydroxylation in vivo with the identification of a new urinary metabolite. See: Cramer, J.W., J.A. Miller, E.C. Miller. 1960. N-Hydroxylation: A new metabolic reaction observed in the rat with the carcinogen 2-acetylaminofluorene. *Journal of Biological Chemistry* 235:885–888.

58 Electrophilic: having an affinity for electrons; pertains to electron acceptance in covalent binding.

59 The relationship of carcinogens to mutagens has been widely discussed. The two groups do NOT coincide exactly but the overlapping of the groups is sufficiently great that, for many investigators, mutagenesis and carcinogenesis become essentially synonymous—to be discussed more fully in chapter 4.

60 Maher, V., E.C. Miller, J.A. Miller, W. Szybalski. 1968. Mutations and decreases in density of transforming DNA produced by derivatives of the carcinogens 2-acetylaminofluorene and *N*-methyl-4-aminoazobenzene. *Molecular Pharmacology* 4:411–426.

major ultimate carcinogen. To complicate understanding of cancer caused by chemicals, it was found that not only are there multiple activated metabolites of a putative carcinogen (e.g., three other enzymatic pathways lead to electrophiles derived from N-hydroxy-AAF) but these active electrophiles are known to combine with multiple nucleophilic[61] sites on DNA. Although the unifying concept of metabolic activation of chemical carcinogens did enhance understanding of chemical carcinogenesis, the obverse occurred with the discovery of multiple chemical metabolites with multiple binding sites on DNA. Cancer biology is indeed uncommonly complex.

In the early 1970s, the Millers investigated metabolic activation of the potent carcinogen aflatoxin B1. Aflatoxin B1 is a product of the common mold *Aspergillus flavus*. The mold and its carcinogenic product are found on grains, nuts and some vegetables, especially where proper storage is unavailable or simply not utilized. At one time it was a major agricultural problem in England, when turkeys were fed mold-contaminated feed. Aflatoxin along with chronic hepatitis B infections may be associated with liver cancer in some African and Asian populations.[62] The Millers reported that aflatoxin B1 becomes mutagenic and carcinogenic after metabolic activation, with a liver preparation that contains active cytochrome P450 monooxygenase or in the intact liver, to form aflatoxin B1 2,3-oxide.[63] These investigations were followed in the Miller lab and by others to show that a major metabolic product of aflatoxin B1 was 2,3-dihydro-2-(guan-7-yl)-3-hydroxyaflatoxin B1[64] which has the competence to bind with DNA.

61 Nucleophilic=an electron donor in covalent bonding.

62 Lin, X., N.J. Robinson, M. Thursz, D.M. Rosenberg, A. Weild, J.M. Pimenta, A.J. Hall. 2005. Chronic hepatitis B virus infection in the Asia-Pacific region and Africa: Review of disease progression. *Journal of Gastroenterology and Hepatology* 20:833–843. See also: Peers, F.G., G.A. Gilman, C.A. Linsell. 1976. Dietary aflatoxins and human liver cancer. A study in Swaziland. *International Journal of Cancer* 17:167–176.

63 Swenson, D.H., J.A. Miller, E.C. Miller. 1975. The reactivity and carcinogenicity of aflatoxin B1-2,3-dichloride, a model for the putative 2,3-oxide metabolite of aflatoxin B1. *Cancer Research* 35:3811–3823.

64 Lin, J.K., J.A. Miller, E.C. Miller. 1977. 2,3-Dihydro-2-(guan-7-yl)-3-hydroxyaflatoxin B1, a major acid hydrolysis product of aflatoxin B1-DNA or -rRNA adducts formed in hepatic microsome-mediated reactions and in rat liver *in vivo*. *Cancer Research* 37:4430–4438. Essigmann, J.M., R.G. Croy, A.M. Nadzan, W.F. Busby, V.N. Reinhold, G. Büchi, G.N. Wogan. 1977. Sturctural identification of the major DNA adduct formed by aflatoxin B1, *in vitro*. *Proceedings of the National Academy of Sciences USA* 74:1870–1874. Smela, M.E., S.S. Currier, E.A. Bailey, J.M. Essigmann. 2001. The chemistry and biology of aflatoxin B1: From mutational spectrometry to carcinogenesis. *Carcinogenesis* 22:535–545.

Liver cancer occurs in rats injected after birth with high doses of safrole, a constituent of sassafras and cinnamon oil flavorings but also found in nutmeg and black pepper. Safrole was used for years as a flavoring ingredient for the soft drink root beer until banned for human food by the United States Food and Drug Administration in 1960. The Millers and their students described the formation of electrophilic metabolites of safrole via the P450-catalyzed reaction. One metabolite, after becoming electrophilic, became a proximate carcinogen deemed to be only one or a few steps from becoming an active carcinogen. The addition of a sulfate group transforms the proximate carcinogen to the ultimate cancer-inducing chemical.[65] And, as in other studies of chemical carcinogenesis, the safrole-derived ultimate carcinogen binds to the electron-donor DNA to produce mutations which results ultimately in uncontrolled; i.e., malignant, growth.

A Synopsis of Metabolic Activation

The non-reactive, complex organic compounds ("procarcinogens") must be metabolically activated[66] to their ultimate reactive carcinogenic form before they react with cellular macromolecules. A clue that metabolic activation of a chemical carcinogen was essential to potentiate its cancer-causing competence was the observation that the cancer often occurred at sites distant from the site of application and regardless of the mode of application. Therefore, one could reason that, if the non-metabolized chemical was intrinsically carcinogenic, cancers should occur at the site of injection, but they did not.

Metabolic activation by cellular enzymes results in the formation of one or more reactive intermediate carcinogen derivatives, termed "*proximate*" carcinogens, which are further metabolized to form one or more *ultimate* electrophilic reactive carcinogens. The Millers were pioneers by their finding of an aminoazo dye metabolite covalently bound to liver proteins in rats.[67] Later, of course, it was shown that the activated chemical covalently bound to RNA and more significantly, DNA. The great majority of chemical carcinogens undergo

65 Miller, J.A., E.C. Miller. 1983. The metabolic activation and nucleic acid adducts of naturally-occurring carcinogens: recent results with ethyl carbamate and the spice flavors safrole and estragole. *British Journal of Cancer* 48:1–15.

66 The widely used term "metabolic activation" was coined by the Millers. Prior to their work, metabolism of a chemical carcinogen was thought to be a step in detoxification.

67 Miller, E.C., J.A. Miller. 1947. The presence and significance of bound aminoazo dyes in the livers of rats fed *p*-dimethylaminoazobenzene. *Cancer Research* 7:468–480.

these chemical reactions to form ultimate carcinogens. An exception to this generalization are the "direct acting" carcinogens which will be considered briefly later in this chapter.

The ultimate carcinogen is a highly reactive electrophile containing electron-deficient atoms (as indicated by the term "electrophile") that seeks to react with nucleophilic sites, i.e., sites that have easily shared electrons. DNA has many nucleophilic sites, and these are critically important in the target cell destined to become malignant. The nucleophilic, electron-donating macromolecules of target cells are, of course, the nucleic acids DNA and RNA as well as protein. However, as James Miller stated "The really critical molecule now appears to be genomic DNA".[68] The covalent adduct, i.e., the macromolecule bound to the activated carcinogen, is the essential material of chemical carcinogenesis. *The Millers were the first to demonstrate that the metabolism of a cancer-causing chemical results in one or more organic compounds with the potential to react with cellular nucleic acids and proteins. Most chemical carcinogens induce cancer primarily by altering critically important nucleic acids, which in turn ultimately results in changed genes (mutation) with altered cells resulting. In short, DNA mutations caused by activated chemical carcinogens is the genesis of cancer.*

The *Essential Material* of Chemical Carcinogenesis

The preceding paragraph identifies the covalent adduct as the "essential material" for the induction of cancer by chemicals.[69] The binding of the activated carcinogen to DNA modifies the DNA such that during replication, RNA transcription is changed and this change results in faulty translation into gene products. This faulty translation does not necessarily result in cancer (and of course, in most cases does not) but if the damage to DNA occurs in one or more genes critical for growth control (e.g., oncogenes or tumor suppressor genes), and if subsequent events occur related to the DNA damage, then cancer may ensue. Of course, tumor formation is a multistep phenomenon the changes of which are the cause and the substance of cancer cell complexity.

---

68 Miller, J.A. 1994. Brief history of chemical carcinogenesis. *Cancer Letters* 83:9–14.

69 See Chapter 12 in: Weinberg, R.A. 2007. *The biology of cancer*. Garland Science (Taylors & Francis). Poirier, M.C., E.K. Weisburger. 2012. Chemical carcinogenesis. In: *Cancer: The Outlaw Cell*, 3rd ed. R.E. LaFond (Ed.), New York: Oxford University Press, pp. 179–188.

Metabolic Activation Requires Enzymes—What Enzymes?

The Millers reported *metabolic destruction*[70] of 4-dimethylaminoazobenzene in intact rats[71] and subsequently that a *metabolic* product of the dye was firmly bound to cell protein.[72] The Millers then decided to follow up their *in vivo*[73] studies with the investigation of the metabolism of carcinogens in vitro using homogenates (also referred to as tissue extracts) of liver.[74]

To prepare the homogenate, the Millers used a homogenizer, which consisted of a glass pestle which fit snugly into an ordinary glass test tube. The pestle was connected to a high speed rotor (1,100 to 1,200 rpm) and the narrow gap between the pestle and the test tube was where tissues were ground and homogenized. I judge that the Millers used the Potter-Elvehjem[75] homogenizer because it had been in existence for several years and was widely used. Further, the scientists who developed the homogenizer, Van Rensselaer Potter (1911–2001) and Conrad A. Elvehjem (1901–1962) were colleagues of the Millers at the University of Wisconsin-Madison.

Homogenize generally means to render uniform throughout. The first glass of a gallon of homogenized milk has the same amount of fat or protein as the last glass. And so correspondingly, one might surmise that a homogenized liver is uniform in every respect but this is not quite true. Actually, fragments of cell structure, such as mitochondria and endoplasmic reticulum, remain intact.

---

70 I use the noun "destruction" in this sentence because the Millers used this term in their original scientific paper and because I believe it conveys a dynamic description of enzymatic activity.

71 Miller, J.A., E.C. Miller, C.A. Baumann. 1945. On the methylation and demethylation of certain carcinogenic azo dyes in the rat. *Cancer Research* 5:162–168.

72 Miller, E.C., J.A. Miller. 1947. The presence and significance of bound aminoazo dyes in the livers of rats fed *p*-dimethylaminoazobenzene. *Cancer Research* 7:468–480.

73 "In vivo" is a term referring to observations made with the use of living animals; "in vitro" refers to studies made using laboratory glassware (which in today's laboratory is generally plastic ware).

74 Mueller, G.C., J.A. Miller. 1948. The metabolism of 4-dimethylaminoazobenzene by rat liver homogenates. *Journal of Biological Chemistry* 176:535–544. The Miller laboratory was the *first* to discover and report oxidative metabolism of a chemical in a cell-free preparation containing enzymes of the cytochrome P450 family of enzymes.

75 Potter, V.R., C.A. Elvehjem. 1936. A modified method for the study of tissue oxidations. *Journal of Biological Chemistry* 114:495–504. This paper describes construction of the Potter-Elvehjem homogeniszer which was not that much different from the Ten Broeck homogenizer used in my laboratory a third of a century later.

The Millers found enzymes (known to be enzymes because they were inactivated by heat and enzyme inhibitors) in the homogenates of whole rat liver. The enzyme system demethylates 4-dimethylaminoazobenzene to 4-monomethylaminoazobenzene and 4-aminoazobenzene.

Earlier it was noted that the carcinogenicity of dimethylaminoazobenzene was modulated by diet, with the less nutritious experimental diet of the Japanese having a greater cancer-promoting competence than the more balanced experimental diet utilized by German investigators. A similar phenomenon emerges from the demethylation studies of liver extracts of mice. Methylation of these compounds is required for carcinogenesis. Less demethylation activity occurred when mice were fed a lower than optimal ground grain or purified diet (thus a diet that enhances carcinogenesis) in contrast to twice the activity of liver homogenates of mice fed a nutritionally better commercial dog chow (with its lower carcinogenesis potency).[76] Diet and carcinogenesis was an early and continuing concern for the Miller laboratory.

A liver homogenate is actually a preparation of cell fragments. Intact nuclei, intact mitochondria and microsomes are present in the homogenate. The homogernate may be described as, to use the German term, a *Gemischt*, i.e., a blend or mixture of cell fragments.

Differential centrifugation was developed as a method of separating the various components of cells in the homogenate. Differential refers to the fact that cellular components settle in the centrifuge at dissimilar rates depending upon their size and density. With gentle centrifugation, intact nuclei, which have a greater size and density than other cellular fragments, sediment to form a pellet. The pellet of nuclei is removed and the remaining material is subjected to greater centrifugal force, which causes mitochondria to settle and form another pellet. Finally, after removal of the mitochondrial pellet, microsomes form a pellet at still higher centrifugal force. What remains are the soluble components in the supernatant fraction, devoid of the previously mentioned particulate cell organelles. Crucial to the findings of the Millers was development of cell fractionation by differential centrifugation with its concomitant understanding of cell structure, biochemistry and function. This

76 Brown, R.R., J.A. Miller, E.C. Miller. 1954. The metabolism of methylated aminoazo dyes. II. Dietary factors enhancing demethylation. *Journal of Biological Chemistry* 209:211–222. Note: The senior author of this paper, Raymond Russell Brown was born in Calgary, Alberta, Canada on 23 December 1926. He graduated from the University of Alberta with a B.S. in 1948. This was followed by a Ph.D. from the University of Wisconsin in 1953, where he became a professor of biochemistry. The composition of the "purified" diet, "grain" and source of the commercial dog chow is given on page 212 of the Brown et al. paper.

technique for the study of cellular components led to a Nobel Prize in 1974 for Albert Claude (1899–1983), Christian René De Duve (1917–2013) and George Emil Palade (1912–2008).

Recall that the Millers reported enzymatic activity in their liver homogenates. Differential centrifugation now afforded the Millers the opportunity to ascribe the enzyme activity to a particular cellular component. The methods of Claude, De Duve, and Palade were probably first used by the Miller laboratory in 1948[77] and they were used again in a careful study to identify intracellular distribution of the enzymes that metabolize *p*-dimethylaminoazobenzene.[78] The azo dye was metabolized by the "small granule" fraction and the soluble (supernatant) fraction. The small granule fraction consisted of microsomes.

If one looks at an electron microscopic image of a cell, one will look in vain for what can be labeled "microsome". Examination of the sequentially formed pellets, homogenized gently and suspended ordinarily in a sucrose solution, reveals that nuclei are intact and look like nuclei. Much the same can be said for mitochondria. However, examination of thin sections of cells with the electron microscope reveal both a granular studded series of membranes, known as the rough endoplasmic reticulum and a series of membranes devoid of the granules, known as the smooth endoplasmic reticulum. The homogenate pellet known as "microsomes" is in fact *fragmented* smooth and rough endoplasmic reticulum, both of which spontaneously form small vesicles. Thus, microsomes, in a sense, are an artifact.[79] Artifact or not, the Miller lab showed quite convincingly that the enzyme activity was in that microsomal fraction composed of fragmented and disrupted endoplasmic reticulum. The microsomal oxidative enzymes, that metabolically activate chemical carcinogens, are located in the smooth endoplasmic reticulum.

---

77 Price, J.M., E.C. Miller, J.A. Miller. 1948. The intracellular distribution of protein, nucleic acids, riboflavin, and protein-bound amino azo dye in the livers of rats fed *p*-dimethylaminoazobenzene. *Journal of Biological Chemistry* 173:345–353.

78 Mueller, G.C., J.A. Miller. 1949. The reductive cleavage of 4-dimethylaminoazobenzene by rat liver: The intracellular distribution of the enzyme system and its requirement for triphosphopyridine nucleotide. *Journal of Biological Chemistry* 180:1125–1136. See also: Brodie, B.B., J. Axelrod, J.R. Cooper, L. Gaudette, B.N. La Du, C. Mitoma, S. Udenfriend. 1955. Detoxification of drugs and other foreign compounds by liver microsomes. *Science* 121:603–604.

79 Artifact (as defined in the *Concise Oxford American Dictionary*, 2006) "something observed in a scientific investigation or experiment that is not naturally present but occurs as a result of the preparation or investigative procedure."—a definition beautifully applicative to the formation of microsomes.

The Paradox of Cancer-Causing Enzymes in Normal Humans

At first thought it seems counterintuitive to contemplate healthy humans metabolically changing benign compounds into highly reactive cancer-causing substances. *Why* do we have enzymes that change the molecular configuration of organic compounds, especially when this may lead to cancer? Someone might suggest that we contain the seeds of our own destruction so that the world does not become over-populated, or that we reap destruction because of our sinful nature, or for whatever notion that comes to mind concerning a built-in intrinsic cancer-causing mechanism in humans. Nonsense! The answer to why we, and all other animals, have enzymes that will metabolically activate potential carcinogens is that we live in what has been termed a sea of xenobiotics.

Xenobiotics are foreign, frequently toxic, lipid-soluble, water-insoluble, non-polar and often large chemical compounds which, when taken into the body, may be toxic. They present a disposal problem. The compounds are potentially toxic because they are non-polar compounds, which are difficult to eliminate because they are not water-soluble (a corollary of being non-polar). With metabolic activation, the compound becomes *more* polar and thereby less competent to traverse the cell membrane to gain access to the interior of the cell, where it would cause havoc. With a water-based urine excretory system, excretion ("disposal") of non-water-soluble foreign compounds is obviously difficult if not impossible. The increased solubility in water results, in most cases, in a reduction in toxicity; hence the term "detoxification" is used for these chemical processes.

Gerald C. Mueller,[80] a Finney-Howell Fellow at the time, and James Miller were the first to show oxygen-dependent metabolism of a xenobiotic[81] in a cell-free preparation of liver enzymes. Their 1948[82] paper showed that enzymes in liver homogenates, "fortified" with diphosphopyridine nucleotide, nicotinamide, magnesium ions and hexose diphosphate, have the potential for demethylation, hydroxylation and reductive cleavage of the azo linkage of 4-dimethylaminoazobenzene. The enzyme activity was subsequently identified

---

80 Gerald Conrad Mueller, born in Centuria, Wisconsin, 1920, obtained his M.D. degree (1943) and his Ph.D. degree (1946) both from the University of Wisconsin-Madison. He became an Assistant Professor of Oncology in 1950 and attained the rank of Professor at the McArdle Laboratory in 1958. He died in 2010.

81 Xenobiotic is a substance (chemical) foreign to a particular organism.

82 Mueller, G.C., J.A. Miller. 1948. The metabolism of 4-dimethylaminoazobenzene by rat liver homogenates. *Journal of Biological Chemistry* 176:535–544.

as an endoplasmic reticulum-associated enzyme of the cytochrome P450 group.[83] The studies of enzymatic conversion of cancer-causing chemicals opened the way for the subsequent study of the cytochrome P450 enzymes that are involved in the metabolism of xenobiotic chemicals, teratogenic chemicals and drugs. There is no better example of the value of basic research.

The P450 complex system of enzymes is coded by a multiple-gene family whose products are involved in oxidative metabolism of foreign chemicals such as sterols, fatty acids, anticancer drugs and a bewildering array of other xenobiotics. The P450 enzymes came into being with the competence to convert (metabolize) xenobiotic compounds into water-soluble substances that are then easily disposed of in urine, bile, feces, breath or perspiration. Most of the time, the metabolic system reduces toxicity and protects humans from the many potentially deleterious xenobiotic chemicals by excretion. Sometimes however, complex organic chemicals (the "procarcinogens") are metabolized, as shown by the Millers, into highly reactive compounds which become, after several metabolic steps, the ultimate carcinogens. The metabolic activation of chemical carcinogens is accomplished by the same enzymes as those used to eliminate xenobiotics. The ultimate carcinogens, produced as we have seen by enzymes which ordinarily function to eliminate xenobiotics, then react with large molecules such as proteins and nucleic acids. Cancer ensues when the ultimate carcinogen and DNA form specific covalent bonds. In brief, this explains how otherwise beneficial enzymes cause molecular changes that result in cancer.

It soon became known that the level of hepatic microsomal enzymes, which metabolize chemical carcinogens, can be increased greatly and rapidly by selected foreign chemicals. This was an important discovery because the enzymes are not only important in activation of chemical carcinogens but also because the enzymes are critical to the metabolism of drugs and environmental chemicals. The phenomenon of increased enzyme production is referred to as "enzyme induction". It was discovered by James Miller and his graduate student Allan H. Conney[84] in the 1950s. Injection of 3-methylcholanthrene,

---

83 Omura, T., R.A. Sato. 1962. A new cytochrome in liver microsomes. *Journal of Biological Chemistry* 237:PC1375-PC1376.

84 Conney, Allan Howard, (1930–2013), was born in Chicago, Illinois. He received his B.S. degree in 1952, M.S. degree in 1954 and his Ph.D. degree in 1956, all from the University of Wisconsin-Madison. He was the Garbe Professor of Cancer and Leukemia Research in the Ernesto Mario School of Pharmacy and Director of the Susan Lehman Cullman Laboratory for Cancer Research at the Rutgers State University, New Brunswick, New Jersey.

benzypyrene and dibenzanthracene induce several-fold increases in the production of hepatic microsomal enzymes.[85] In addition to those injected chemicals, more than 200 enzyme inducers were known as long ago as 1967[86] including certain hypnotics, sedatives, anesthetic gases, central nervous system stimulators, anticonvulsants, tranquilizers, antipsychotics, hypoglycemic agents, anti-inflammatory agents, muscle relaxants, antihistamines, alkaloids, insecticides, food additives, steroid hormones, industrial chemicals and certain carcinogens.

Enzyme induction is of profound significance because it results in enhanced carcinogenicity of chemicals that are ordinarily metabolically activated by these enzymes, which of course is the primary subject of the Millers' work. Moreover, enzyme induction increases chemical transformation of drugs and environmental chemicals with the results of altered pharmacologic and toxicologic properties. With regard to drug metabolism, the biotransformation alters duration and intensity of drugs in humans.

As indicated, enzyme induction is a topic of critical importance not only for carcinogen metabolism and identification (several rapid tests for carcinogenic potential are described in the next chapter) but also to pharmacology (drug metabolism)—and the Miller lab was the *first* to discover and study it.

Charles Darwin would argue that traits that are detrimental to the species are not retained. This is referred to as natural selection. One could surmise that the enzyme system that results in metabolic activation of potential carcinogens must, therefore, enhance survival of the species. How odd! A substance that causes cancer enhances survivability? Since we live in the sea of xenobiotics alluded to earlier, the xenobiotics in my opinion must be of *greater* hazard than our encounters with potential carcinogens. The ubiquitous presence of xenobiotic metabolizing enzymes throughout the animal world witnesses to this fact. Added to this argument, one could reason that because many, if not most, of the carcinogens that humans encounter in their environment derive from human activity, an evolutionary process has not yet yielded a means of

---

85 Conney, A.H., E.C. Miller, J.A. Miller. 1956. The metabolism of methylated aminoazo dyes. v. Evidence for induction of enzyme synthesis in the rat by 3-methylcholanthrene. *Cancer Research* 16:450–459. Conney, A.H., E.C. Miller, J.A. Miller. 1957. Substrate-induced synthesis and other properties of benzpyrene hydroxylase in rat liver. *Journal of Biological Chemistry* 228:753–766. Conney, A.H., J.R. Gillette, J.K. Inscoe, E.R. Trams, H.S. Posner. 1959. Induced synthesis of liver microsomal enzymes which metabolize foreign compounds. *Science* 130:1478–1479. See also: Snyder, R. 2000. Microsomal enzyme induction. *Toxicological Sciences* 55:233–234.

86 Conney, A.H. 1967. Pharmacological implications of microsome enzyme induction. *Pharmacological Reviews* 19:317–366.

coping with human-made chemical carcinogens. Thus, for our time in the evolution of protective mechanisms, we must cope with an imperfect P450 protective system.

One could add, whilst considering Darwinian evolution and the "imperfect P450 protective system", that most cancer in humans occurs *after* reproduction, and thus has essentially no effect on fitness (if fitness is defined in reproductive terms). This notion of fitness and age at reproduction provides support for the question of why we harbor an enzyme system capable of activating chemical carcinogens.

Direct Acting Chemical Carcinogens

Diverse paths may lead to a commonality. The "diverse paths" refers to the requirement for either metabolic activation, or its lack, with regard to chemicals that cause cancer. The "commonality" is the resulting cancer regardless of the particular etiological "path(s)". In its simplest form, the dictum requires that chemical carcinogens must react with the DNA of the host with the formation of covalent adducts. Cancer ensues under certain circumstances. The chemicals not requiring metabolic activation are referred to as "direct acting" carcinogens and they include ethyleneimine, ß-propriolactone, bis (2-cholorl-ethyl) sulfide, nitrogen mustard, bis (chloromethyl) ether, diepoxybutane, ethylene oxide, and methyl methane sulfonate.[87]

The direct-acting carcinogens, also known as activation independent carcinogens, are characterized by inherent reactivity that needs no metabolic conversion to form adducts with DNA, RNA and proteins. As a group, they tend to be less hazardous than chemicals requiring metabolic activation simply because they are very reactive and generally interact with cellular constituents *prior* to reaching their "target" of DNA. As such, they lose much of their cancer-causing potential resulting in minimal genetic damage. However, in contrast to the much more numerous chemical carcinogens that require metabolic activation, direct acting chemical carcinogens may cause cancer at the site of exposure. Several chemotherapeutic agents belong to the direct acting group and the dismal potential, in a limited number of cases, of causing a second cancer

87 Phillips, S.D., M.I. Greenberg, M.T. Paria. 2003. Chemical carcinogenesis. In: Greenberg (ed.), *Occupational, Industrial, and Environmental Toxicology*, 2nd Ed. Philadelphia: Mosby Inc., pp. 670–681. Pitot, H. 2002. *Fundamentals of Oncology*, 4th edition, New York: Marcel Dekker, Inc.

after cure of the treated malignancy. Direct acting chemical carcinogens were not the primary interest of Elizabeth and James Miller.

Of course, carcinogen/DNA binding is only a first step. I call attention to the epigraph at the beginning of this chapter (National Academy of Sciences, 2009). Binding of an activated carcinogen to specific sites on DNA may result in mutation to proto-oncogenes and/or tumor suppressor genes. Those specific mutations will lead to changes in gene expression which in turn may lead to cancer.

Proto-Oncogenes and Tumor Suppressor Genes

As we have learned in this chapter, the principle focus of the Millers' research showed that most chemical carcinogens are activated by enzymes to become reactive electrophiles that bind covalently to nucleophilic sites on DNA. These reactions lead to a form of mutagenesis. Mutations of *random* genes is unlikely to lead to cancer (dysfunctional cells, premature cell death or other cell abnormalities may occur but cancer is not particularly likely). A rhetorical question follows: If not *random* genes, then mutation to which genetic sites (loci) leads to cancer?

Proto-oncogenes and tumor suppressor genes are significant targets for mutation by chemical carcinogens. Genetic changes in those genes play a crucial role in the genesis of cancer.[88] Proto-oncogenes , in normal cells, encode for proteins that stimulate growth, retard cell differentiation and impede cell death; they promote *normal* growth and proliferation. The mutated version of the gene, now referred to as an oncogene, results in an increased productivity of these controlling proteins, the over production of which results in excess cell division, inhibited cell differentiation and blocked apoptosis ("immortalization)—together these attributes are *early hallmarks of cancer*. Tumor suppressor genes are normal genes which function by elaboration of proteins that restrain cell division. The mutation of tumor suppressor genes result in the loss of that restraint, i.e., cell division will increase substantially and abnormally. Proto-oncogenes and tumor suppressor genes are only a small fraction of the activated carcinogen binding sites but, as described here, they are "significant"

88 Hanks, S.K., S.R. Hann. 2012. Oncogenes. In: R.E. LaFond (ed.), *Cancer: The Outlaw Cell*, 3rd ed., Oxford University Press, New York, pp. 27–52. Frank, A., J. Pimkina, M.E. Murphy. 2012. Tumor Suppressor Genes, In: R.E. LaFond (Ed.) *Cancer: The Outlaw Cell*, 3rd ed., Oxford University Press, New York, pp. 67–76.

sites and have been described elsewhere as "critical genomic sites" for the induction of cancer.

Hecht summarized a huge amount of scientific literature into two short sentences:[89] "If a permanent mutation occurs in a critical region of an oncogene or tumor suppressor gene, it can lead to activation of an oncogene or deactivation of the tumor suppressor gene. Multiple events of this type lead to aberrant cells with loss of normal growth control and ultimately to lung cancer." Hecht went on to write: "cancer induction is a process which requires multiple genetic changes." In a similar manner, Stanley wrote "a multitude of molecular changes occur during tumorigenesis."[90]

The reason that the phrase "*early hallmarks of cancer*" in a previous paragraph was italicized relates to the notion that the genesis of cancer cannot be related to one or two genetic events. A multiplicity of changes in genetics and cell function occur with cancer, and mutated proto-oncogenes and tumor suppressor genes are but a few of the early changes. Cancer research is never easy—the "multiplicity of changes" include those which are an effect rather than a cause of transformation and, of course, the molecular biologist must eventually sort out effect from causes.

The Significance of the Millers' Research

"James A. Miller and Elizabeth C. Miller discovered and developed the important unifying concept that most carcinogenic and mutagenic chemicals are not carcinogenic or mutagenic *per se*, but that those compounds must undergo metabolic transformation to reactive electrophilic metabolites that manifest their effects by covalently binding to critical sites on cellular macromolecules (DNA, RNA, and protein)".[91] The actual development of the malignant state follows by alteration of specific proto-oncogenes and/or tumor suppressor genes. Their endeavors revealed the earliest steps in carcinogenesis, they fused carcinogenesis with mutagenesis providing new insight into the ontogeny of cancer and their studies enabled tests for the detection of cancer-causing chemicals.

---

89 Hecht, S.S. 1999. Tobacco smoke carcinogens and lung cancer. *Journal of the National Cancer Institute* 91:194–1210.

90 Stanley, L.A. 1995. Molecular aspects of chemical carcinogenesis: The roles of oncogenes and tumor suppressor genes. *Toxicology* 96:173–194.

91 Conney, A.H., M.C. Poirier, Y.-J. Surh, F.F. Kadlubar. 2009. *Elizabeth Cavert Miller 1920–1987 James A. Miller, 1915–2000, A Biographical Memoir.* Washington, D.C.: National Academy of Sciences.

Other areas of research are considered in the next chapter—altogether, their contributions elevated the Millers, in the judgment of many wise individuals, to the highest level of intellectual endeavors. The endeavors had the potential for, and the realization of, procedures for sparing multitudes of human lives.

The Quest Continues

Henry Pitot, a colleague of the Millers and director emeritus of the McArdle Laboratory for Cancer Research, stated that while much has been learned about chemical carcinogenesis in the past half century or so, "we still seem to be a long way" from the understanding of how cancer occurs after exposure to a chemical carcinogen.[92] The Millers focused on metabolism of these chemicals to ascertain the ultimate active form(s) as the first step in unraveling the mechanism of how chemicals cause cancer. Earlier, they put it this way: "The only certain indicator that the carcinogenic process has occurred or has been in progress in a given tissue is the eventual emergence of a gross tumor.[93] Consequently the task of defining the carcinogenic process is still largely restricted to the correlation of events in the target tissue with the ultimate appearance of gross tumors."[94] These truths with little change have kept scientists such as Elizabeth and James Miller at the laboratory bench. As each problem is "solved", many more are brought to light and the scientific quest continues. A question impossible to answer but deserving thoughtful contemplation is how many lives have been saved and will be saved due to the research of the Millers?

---

92 Pitot, H.C. 2002. *Fundamentals of Oncology*, 4th edition. New York: Marcel Dekker, Inc., page 49.

93 The leukemias do not present as "gross tumors". Nevertheless, the search for early events in the target tissue, i.e., hematopoietic stem cells, may serve a role in defining leukemogenesis.

94 Miller, J.A., E.C. Miller. 1953. The carcinogenic aminoazo dyes. *Advances in Cancer Research* 1:339–396.

CHAPTER 4

Serendipity: How the Millers Unintentionally Revolutionized Biology

Science is never static. Science is always moving forward. The rate of that forward movement and the extent of advancement in various allied fields are a measure of the importance of that particular branch of knowledge. Such is the case with the life-time research of Jim and Betty Miller. The Millers' research was the scientific foundation of "(1) rapid mutagenicity tests (the Ames test and others) for the screening of potential human carcinogens, (2) research on chemically induced mutation in proto-oncogenes and tumor suppressor genes, (3) recognition of large person-to-person differences in the sensitivity of people to environmental carcinogens (intraindividual differences in the metabolism of carcinogens to their active forms), (4) identification of macromolecular adducts as indicators of cancer risk in molecular epidemiology, and (5) international laws regulating carcinogens in the diet and the environment."[1] With the exception of chemically induced mutations of proto-oncogenes and tumor suppressor genes which are discussed in the previous chapter, these domains of modern toxicology are discussed below.

A Note about Basic Research

Because the Millers' cancer studies were *basic research* and this chapter considers research engendered by that *basic research*, it seems appropriate to consider the value of basic research in general and that of the Millers in particular.[2] I agree with Blakemore and Davidson[3] who stated, "Trust in research requires a leap of faith. This is especially the case for basic research not motivated by the immediate prospect of application. *But such research is the seed from which new treatments grow*" (italics added). Substitute "*new understanding*" in lieu of

1 See epigraph of Chapter 3.

2 See: Curiosity Creates Cures: *The Value and Impact of Basic Research*, 2011. http://progressreport.cancer.gov/sites/default/files/archive/report2011.pdf.

3 Blakemore, C., J. Davidson. 2006. Putting a value on medical research. *Lancet* 367:1293–1295.

 | DOI 10.1163/9789004286801_005

"*new treatments*" in Blakemore and Davidson's metaphor, and we have a measure of the value of the many related studies brought fourth by the Millers.

Several years ago, there was a symposium on medical research and practice held in London. C.J. Dickinson, of Saint Bartholomew's Hospital, London and one of the symposium participants, endeavored to evaluate the relative value of basic medical research versus clinical research.[4] This, as it might be imagined, provoked considerable discussion including that of H.H. Hiatt of Harvard University. Hiatt cited the work of Bruce Ames "who has recently developed one of the most important assays for chemical mutagenesis, and, inferentially, of carcinogenesis. Ames's brilliant studies of the control of histidine[5] biosynthesis began in the mid-50s. But, understandably, none of his early fundamental work was financed by cancer research funds, for neither he nor anybody else could have suspected the potential applicability of his work to cancer studies."[6] The Hiatt quotation is pertinent here for two reasons: (1) Hiatt's statement relates to the difficulty of predicting usefulness of basic research and (2) the Ames' procedure for detecting mutagenesis *absolutely depended* upon prior research of the Millers. Clearly, the mutagenesis assay of potential carcinogens would not exist without the enzyme activity discovered and described in rat liver homogenates by Elizabeth and James Miller with their coworkers.

Another participant at the London Ciba Symposium, M.F. Perutz of Cambridge, U.K., also speaking of the Ames assay, recognized the importance of the Millers' research. Perutz stated that many compounds are "non-carcinogenic as such, but are converted into carcinogens" by enzymes in the liver. Perutz then reported that Ames "decided to mimic this conversion by adding rat or human liver homogenate" to the test system, thereby detecting almost all of the carcinogens that required metabolic activation. The Millers' research was *the seed* that enabled the development of the Ames assay, and other tests, for mutagenesis.[7] Those tests have the important potential of cancer prevention.

---

4 Dickinson, C.J., 1976. The value of medical practice to basic and applied medical research done twenty years ago. In: *Research and Medical Practice: their interaction.* Ciba Foundation Symposium 44, Amsterdam, Elsevier, pp. 53–71.

5 Histidine is an alpha amino acid.

6 Hiat, H.H. 1976. Comment to discussion of C.J. Dickinson's paper, see reference in note 4, page 66.

7 The in vitro study of the Miller laboratory was the first to show metabolism of a chemical carcinogen to an activated state in a cell-free enzyme. See: Gelboin, H.V., J.A. Miller, E.C. Miller. 1958. The formation in vitro of protein-bound derivatives of aminoazo dyes by rat liver and its enhancement by benzpyrene. *Biochemica et Biophysica Acta* 27:655–656. This became the foundation of the subsequent Ames test for mutagens.

Elizabeth Miller, in her presidential address to the American Association for Cancer Research,[8] stated "the generalization that strong electrophilic reactivity is a basic requirement for ultimate chemical carcinogens and detailed knowledge of enzyme systems involved in the metabolic activation and deactivation of many chemicals *are providing useful tools for the recognition of chemicals to be suspected of being potential carcinogens*" (italics added). Methods for the recognition of potential carcinogens are discussed below.

Test Systems for Mutagens and Carcinogens

Early in the task of writing this book, a friend of mine asked what "metabolic activation of chemical carcinogens" (i.e., the lifetime work of Betty and Jim Miller) had to do with the understanding, prevention or control of human cancer. My friend had a somewhat cynical view that if a cancer cure was not being pursued, then the basic research was an activity of little value; it could be considered to be just so much "busywork". Since the writing of this book has taken several years, I can now affirm that I have thought about that question for some time. I also gave thought to the meaning of "basic research". It is worth repeating here the statement of Dr. Henry C. Pitot, who wrote that the research of the Millers "... may prove to be the most important of this century in our understanding, prevention and control of human cancer."[9] Clearly, tests that detect potential carcinogens certainly relate to the understanding, prevention and control of human cancer—this is exactly the issue to which my friend inquired.

Some chemicals cause cancer. Which chemicals? It is extraordinarily expensive to expose a large number of experimental animals to a new chemical that might be carcinogenic and wait months to see if a cancer appears. Technician labor is expensive, animal testing is labor intensive, funds are limited, and many new chemicals appear each year. As long ago as 1980, it was estimated that animal testing for a new chemical could cost as much as £500,000 in England. With inflation, the figure for 2015 would amount to about £2,250,000!—or, about $3,300,000 in US Dollars. Remember that this is for carcinogenesis testing of only *one* chemical—multiply the cost in £ or $ by 80,000, the number of new chemicals introduced into our society each year, and clearly a massive and unaffordable figure emerges. Why is the cost so high? The cost of animal

8 Miller, E.C. 1978. Some current perspectives on chemical carcinogenesis in humans and experimental animals: Presidential address. *Cancer Research* 36:1479–1496.

9 See Chapter One.

testing relates to the numbers of animals used and to such complex factors as age, sex, species, quality and quantity of diet, dosage schedule, route of administration, and vehicle or solvent used, as well as the cost of record keeping and animal caretakers. Carcinogenesis testing is far more complex than simply feeding a potential carcinogen in a single dose of one concentration in one solvent to one kind of animal.[10]

There was (and continues to be) a need for faster and cheaper tests. Most tests are based upon detecting genetic change in an organism after exposure to a chemical. There is a long history relating genetic change to cancer.[11] This history evolved from identification of aberrant chromosomes found in cancer[12] to seeking more subtle changes in cancer cell gene expression.[13] M.-C. King and her colleagues put it this way: "For decades there has been no doubt that cancer is genetic, in the sense that transformation of a normal cell to invasive and malignant growth is due to *changes in the* DNA (italics added)."[14] A definition of mutation is a change or alteration in DNA. Thus, if genetic changes in a cell are a hallmark of cancer, then it would prove useful to explore how to detect that genetic (mutational) change.[15] It has been recognized for many years that the frequency of mutation in humans is considered to be a serious hazard and for this reason, the testing of chemicals for mutagenic activity before they enter the human environment is an absolute high priority.

Biologists used the fruit fly, *Drosophila melanogaster*, as the "most important model organism for genetic studies since early in the twentieth century."[16]

---

10 Boyland, E. 1980. The history and future of chemical carcinogenesis. *British Medical Bulletin* 36:5–10.

11 Boveri, T. 1914. *Zur Frage der Erstehung Maligner Tumoren*. Jena: Gustav Fischer. (English translation): Boveri, M.O. 1929. *The Origin of Malignant Tumors*. Baltimore: Williams and Wilkins Company.

12 Mitelman F. 1994. *Catalog of chromosome aberrations in cancer*, 5th edition. New York: Wiley-Liss.

13 Macgregor, P.F., J.A. Squire. 2002. Application of microarrays to the analysis of gene expression in cancer. *Clinical Chemistry* 48:1170–1177.

14 King, M-C., S. Rowell, S.M. Love. 1993. Inherited breast and ovarian cancer. What are the risks? What are the choices? JAMA 269:1975–1980.

15 It is now known that there are also epigenetic changes that are heritable and can result in cancer. These changes include methylation of DNA and histone acetylation. See: Hassler, M.R., G. Egger. 2012. Epigenomics of cancer—emerging new concepts. *Biochimie* 94:2219–2230; Dawson, M.A., T. Kouzarides. 2012. Cancer epigenetics from mechanism to therapy. *Cell* 150:12–27.

16 Snustad, D.P., M.J. Simmons. 2012. *Principles of Genetics*, 6th edition. Hoboken, N.J.: John Wiley & Sons, Inc.

The "high level of gene and pathway conservation, the similarity of cellular processes and the emerging evidence of functional conservation of tumor suppressors between *Drosophila* and mammals"[17] suggest that the fruit fly should be an important test organism. It is emerging as such but *Drosophila* has limitations discussed elsewhere.[18] While there are multiple assays for chemical carcinogens,[19] this discussion will focus on a bacterial assay and assays in vertebrates. The tests that will be described here were developed at the time the Miller laboratory was active and depended upon the capacity of the test system to metabolically activate potential carcinogens to become not only cancer-causing chemicals but chemicals with mutagenic potential as well.[20]

Genetic toxicology was rapidly evolving concurrently with the Millers' research. It is no surprise, then, that investigators considered screening for mutagens by incorporating a rodent's ability to metabolically activate a putative mutagen. A truly novel procedure was introduced by Marvin Legator,[21] who used mice because of their competence for metabolic activation. The indicator organism for mutation in the Legator procedure was a strain of *Salmonella typhimurium* which was injected into the recipient mouse along with the chemical to be studied. The reason for the mouse as a host is that bacteria in general, and *Salmonella* specifically, do not have the carcinogen-metabolizing enzymes found in the cells of the mouse (and other vertebrates including humans). The mouse liver performs the metabolic activation of the mutagen and the *Salmonella* responds to the metabolic products of the tested chemical

17 Potter, C.J., G.S. Turenchalk, T. Xu. 2000. Drosophila in cancer research: an expanding role. *Trends in Genetics* 16:33–39. See also: Vidal, M., R.L. Cagan. 2006. Drosophila models for cancer research. *Current Opinion Genetics & Development* 16:10–16.

18 Giacomotto, J., L. Ségalet. 2010. High-throughput screening and small animal models, where are we? Review. *British Journal of Pharmacology* 160:204–216.

19 Montesano, E.R., H. Bartsch, L. Tomatis (eds.), *Screening tests in chemical carcinogenesis*. Lyon: IARC Scientific Publications No. 12. Maxwell, C.K., P.L. Leung, A. Bennedetto, C. Au, K.J. Helmcke, M. Aschner, J.N. Meyer. 2008. *Caenorhabditis elegans*: An emerging model in biomedical and environmental toxicology. *Toxicological Sciences* 106:5–28.

20 Maher, V.M., E.C. Miller, J.A. Miller, W. Szbalski. 1968. Mutations and decreases in density of transforming DNA produced by derivatives of the carcinogen 2-acetyl-aminofluorene and N-methyl-4-aminoazobenzene. *Molecular Pharmacology* 4:411–426. Miller, E.C., J.A. Miller. 1971. The mutagenicity of chemical carcinogens: Correlations, problems, and interpretations. In: *Chemical Mutagens—Principles and Methods for their Detection*. Vol. 1. A. Hollaender (ed.), New York: Plenum Press, pp. 83–119.

21 Marvin S. Legator (1926–2005), was born in Chicago, he received his Ph.D. from the University of Illinois, and was a Professor at the University of Texas Medical Branch, Galveston, Texas.

while in the host's body. After this treatment, *Salmonella* are recovered from the mouse and cultured *in vitro*. Mutations in the test culture are then scored and compared with control cultures of *Salmonella*.[22] The test is referred to as "host-mediated" assay, which means that the host (the mouse) provides the enzymes which mediate (i.e., bring about the result of) the metabolic activation. Because of the already perceived notion that mutagens are also carcinogens, the host-mediated assay becomes a carcinogen test that detects mutation and bypasses tumor formation in the host rodent—a money and time-saver indeed. Gabridge and Legator knew of the metabolizing competence of mammalian enzymes and appropriately credited Elizabeth C. and James A. Miller.

A refinement of the host-mediated assay does away with the intact and alive mouse but uses the mouse's liver for the preparation of microsomes. Recall that the Millers showed that the enzymes necessary for metabolic activation were in the microsome fraction of liver. Early attempts to use rodent microsome fractions for studies in mutagenesis and toxicology were reported by Malling in 1971[23] and the Millers with their colleagues Garner, Garner and Hanson, also in 1971.[24] The latter group reported that aflatoxin became lethal to *Salmonella typhimurium* after incubated with liver microsomes (as an aside, the Miller group made certain that their audience knew exactly what the aflatoxin was incubated with; it was termed "rat liver post-mitochondrial supernatant").

Later Bruce N. Ames[25] developed the well-known and widely used test[26] which is an in vitro *Salmonella*-microsome[27] assay that combines the indicator

---

22 Gabridge, M.G., M.S. Legator. 1969. A host-mediated microbial assay for detection of mutagenic compounds. *Proceedings of the Society for Experimental Biology and Medicine* 130:831–834. Legator, M.S., T. Conner, M. Stoeckel. 1975. The detection of mutagenic substances in the urine and blood of man. *Annals of the New York Academy of Sciences* 269:16–20.

23 Malling, H.V. 1971. Dimethylnitrosamine: formation of mutagenic compounds by interaction with mouse liver microsomes. *Mutation Research* 13:425–429.

24 Garner, R., E.C. Miller, J.A. Miller, J.V. Garner, R.S. Hanson. 1971. Formation of a factor lethal for *S. typhimurium* TA1530 and TA1531 on incubation of aflatoxin B1 with rat liver microsomes. *Biochemical and Biophysical Research Communications* 45:774–780.

25 Ames, Bruce N (born 1928) is a Professor of Biochemistry and Molecular Biology at the University of California Berkeley. This is the same Bruce Ames referred to several times in the 1976 Ciba Foundation meeting in London referred to previously.

26 Ames, B.N., W.E. Durston, E. Yamasaki, F.D. Lee. 1973. Carcinogens are mutagens: A simple test system combining liver homogenates for activation and bacteria for detection. *Proceedings of the National Academy of Sciences USA* 70:2281–2285. Also see: Venitt, S. 1980. Bacterial mutation as an indicator of carcinogenicity. *British Medical Bulletin* 36:57–62.

27 Perhaps a better but admittedly longer designation would be "liver microsome mixed-function oxygenase system."

organism *Salmonella* with an in vitro mammalian enzyme fraction to test for mutagenesis. Scrutiny of chemical carcinogens with the Ames test revealed that most carcinogens (90%) were mutagenic.[28] More recently, the bacterial strains used in mutagenesis testing have been engineered to express human enzymes, which enhances the potency of the assay.[29] A historian of science would have no difficulty in relating the development of the Ames test to the pre-existing metabolic studies of the Millers.[30]

A number of years ago, I was working with a huge chemical company to design a quick and reliable test for chemical mutagenesis. The company was using cytogenetic (chromosomal) analysis for screening their employees. The rationale was to detect early on if the manufacturing process was leading to chromosomal abnormalities among its workers. Increased chromosomal abnormalities would be a "red flag" indicating potential hazard in some step in the production process. Presumably, detecting a potential mutagen or carcinogen in an animal test would be quicker, cheaper, and much more humane.

The test under development used living animals, with intact livers, and thus would depend upon the innate competence of the animal to metabolically activate the putative mutagen. It was hoped that the assay would be more sensitive and dependable than the microbiological tests previously described. The North American Leopard Frog, *Lithobates pipiens,* (until recently and better known as *Rana pipiens*), was chosen as the experimental animal[31] because frogs are vertebrates and share biochemical mechanisms with humans, and because the frog's reproductive system was well-known and easy to manage. Furthermore, the ubiquitous Leopard Frog, widely distributed in much of the United States and contiguous parts of Canada, was abundant and relatively cheap. A long ago colleague of mine, Dante J. Picciano,[32] my students and I sought to detect subtle changes in the frog DNA after exposure to a mutagen.

28 McCann, J., B.N. Ames. 1976. Detection of carcinogens as mutagens in the *Salmonella*/ microsome test: Assay of 300 chemicals: Discussion. *Proceedings of the National Academy of Sciences USA* 73:950–954.

29 Josephy, P.D. 2002. Genetically-engineered bacteria expressing human enzymes and their use in the study of mutagens and mutagenesis. *Toxicology* 181–182:255–260.

30 See: Zeiger, E. 2004. History and rationale of genetic testing: An impersonal, and sometimes personal, view. *Environmental and Molecular Mutagenesis* 44:363–371.

31 McKinnell, R.G., D.J. Picciano, J.W. Schaad IV. 1979. Dominant lethality in frog embryos after paternal treatment with triethylenemelamine. *Environmental Mutagenesis* 1:221–231.

32 Picciano, Dante J., born 1943 in Palmerton, Pennsylvania. He received his Ph.D. from George Washington University in 1973 and a J.D. from Capital University School of Law in 1985. He is currently a lawyer in Tamaqua, Pennsylvania.

What was absolutely essential for the test was the necessity of frogs sharing with humans the competence to metabolically activate potential mutagens. In point of fact, the frogs do indeed have that metabolic competence,[33] which illustrates, among other things, the antiquity of this particular metabolic competence (frogs have been around since the Jurassic Period, about 150 to 200 million years ago; Jurassic frogs were contemporaries of dinosaurs) and that present-day frogs have metabolic capability relating to survival comparable to that of humans. Naive indeed were we to have wondered if frogs had the competence to metabolically alter xenobiotics. How would frogs survive in the environment that we have if they did not possess the biological capability to metabolize and excrete the end products of xenobiotics? The frog test did not emerge as a widely used procedure—but, in retrospect, it highlights the universal need for protection from xenobiotics. Further, the simultaneous finding that dose related chromosomal aberrations go hand in hand with dose related morphological abnormalities in the testing of a single putative carcinogen reveals the close relationship of teratogenesis with mutagenesis and thus, carcinogenesis.

Mutations in proto-oncogenes and tumor suppressor genes were discussed in the previous chapter.

Person to Person Differences in Sensitivity to Environmental Carcinogens

In the first paragraph of his 1989 paper, Curtis C. Harris[34] quoted Charles Robert Darwin who wrote "No one supposes that all the individuals of the same species are cast in the very same mould. These individual differences are highly important for us..."[35] Allan H. Conney (Figure 4.1) asked in his Clowes Memorial Lecture "Why does a drug or an environmental pollutant cause toxicity in one person and not in another person? Why do some cigarette smokers develop lung cancer whereas other cigarette smokers do not? One of the

33 Cheh, A.M., A.B. Hooper, J. Skochdopole, C.A. Henke, R.G. McKinnell. 1980. A comparison of the ability of frog and rat S-9 to activate promutagens in the Ames test. *Environmental Mutagenesis* 2:487–508.

34 Harris, C.C. 1989. Interindividual variation among humans in carcinogen metabolism, DNA adduct formation and DNA repair. *Carcinogenesis* 10:1563–1566.

35 Darwin, C.R. 1861. *The Origin of Species*, 3rd edition with additions and corrections. London: John Murray, 50 Albemarle Street, Mayfair. Chapter II. Variation under nature. page 47.

FIGURE 4.1 *Allan H. Conney, a former student and colleague of the Millers, questioned why a drug or an environmental substance caused toxicity in one individual but not in another.*
PHOTOGRAPH BY MICHAEL DURINZI. COURTESY OF JOSEPH BARONE, ERNEST MARIO SCHOOL OF PHARMACY, RUTGERS, THE STATE UNIVERSITY OF NEW JERSEY.

causes of variation in the response of human beings to foreign chemicals is dissimilarity in the rate of metabolism of those chemicals."[36] Variations in sensitivity to chemical carcinogenesis are dependent upon race, age, developmental stage, sex, occupation, nutritional condition and additional factors. In the

36 Conney, A.H. 1982. Induction of microsomal enzymes by foreign chemicals and carcinogenesis by polycyclic aromatic hydrocarbons: G.H.A. Clowes Memorial Lecture. *Cancer Research* 42:4875–4917. Conney and the Millers were on the cover of *Cancer Research*, Vol. 50, March 15, 1990 (see Appendix).

context of this chapter, the individual differences in sensitivity are important and in part, explain the great variation in cancer vulnerability among individuals. Cancer generally follows damage to DNA but, as indicated here, individual variation in cancer vulnerability depends upon enzymes that metabolically activate most carcinogens (see discussion of the cytochrome P450 enzymes in preceding chapter).

In this regard, consider the cigarette smokers referred to by Conney. Most lung cancer occurs among cigarette smokers. However, most cigarette smokers *do not* succumb to lung cancer.[37] Almost everyone knows of someone who has been a heavy smoker for nearly a lifetime but who at a ripe old age dies of something not related to smoking—such as an accidental gun shot while hunting or by walking in front of a moving bus.

It was written in a stimulating scientific paper that "Historically, policy makers such as the U.S. Environmental Protection Agency have based their decisions on the assumption that all individuals in a population have the same biologic response to a specified dose of a carcinogen."[38] That assumption is incorrect but incorrect in what way? Divergence is found in the enzyme activity that metabolizes carcinogens which, in turn, will form adducts to DNA.[39]

37 In a Canadian study, the lifetime risk for lung cancer in a male smoker was reported to be 17.2%. A European study reported the lifetime risk of ordinary male smokers to be 15.9% but "heavy smokers" had a lifetime risk of 24.4%. Female smokers had similar but lower risks. See web site of Lynne Eldridge: http://lungcancer.about.com/od/Lung-Cancer-And-Smoking/f/smokers. A question posed in this chapter and this section is how do the majority of lifetime smokers escape lung cancer? Insight to this quandary, but no explanation, derives from the fact that tobacco is not a *necessary* cause for lung cancer; i.e., some non-smokers succumb to lung cancer, and tobacco is not a *sufficient* cause for cancer as evidenced by the fact that most smokers do not die of lung cancer.

38 Perera, F.P. 1997. Environment and cancer: Who are susceptible? *Science* 278:1068–1073.

39 Shimada, T., H. Yamazaki, M. Mimiura, Y. Inui, F.P. Guengerich. 1994. Intraindividual variations in human liver cytochrome P450 enzymes involved in the oxidation of drugs, carcinogens and toxic chemicals: Studies with liver microsomes of 30 Japanese and 30 Causasians. *The Journal of Pharmacology and Experimental Therapeutics 270:414–264.* Sin, D.D., S. Ben-Zaken Cohen, A. Day, H. Coxson, P.D. Paré. 2007. Understanding the biological differences in susceptibility to chronic obstructive pulmonary disease between men and women. *Proceedings of the American Thoracic Society* 4:671–674. Ben-Zaken Cohen, S., P.D. Paré, S.F. Man, D.D. Sim. 2007.The growing burden of chronic obstructive pulmonary disease and lung cancer in women: examining sex differences in cigarette smoke metabolism. *American Journal of Respiration and Critical Care Medicine* 176:113–120. Oyama, T., K. Sugio, T. Isse, A. Matsumoto, A.N. Nose, H. Uramoto, T. Nozoe, M. Morita, N. Kagawa, T. Osaki, M. Muto, K. Yasumoto, T. Kawamoto. 2008. Expression of Cytochrome

This variation in enzyme activity, in part, explains why diversity, not uniformity, in prevalence of specific cancers occurs. Please consider the well know fact that more elderly than young people die of cancer.[40] There are multiple explanations for this diversity including the obvious fact that elderly individuals are exposed to cancer-causing agents longer than the young, i.e., the *duration* of exposure to most carcinogens is greater among the aged.

In a perhaps more subtle examination of age-related diversity, it was shown in mice that senescent animal livers yielded enzymes that produced a greater yield of activated chemical carcinogen metabolites than do the livers of younger animals. Further, the activated metabolites produced in the older mice yielded a higher level of covalent binding to host DNA than did the activated metabolites from younger animals.[41] While these studies suggested the possibility that some portion of the increased prevalence of age-related cancer in mice may be attributable to increases in carcinogen metabolism, other studies with rats report a decrease or unchanged level of hepatic enzyme activity.[42]

The carcinogen-metabolizing enzymes were, of course, characterized first in the laboratory of Elizabeth and James Miller. The Millers' work provides an explanation of why some people do, and others do not, succumb to cancer.

Molecular Epidemiology: Formation of Macromolecular Adducts as Indicators of Cancer Risk

Epidemiology is the study of prevalence and geographic distribution of disease with the ultimate goal of disease prevention. Cholera is an example of an

---

P450 in non-small cell lung cancer. *Frontiers in Bioscience* 13:5787–5793. Anttila, S., H. Raunio, J. Hakkola. 2011. Cytochrome P450-mediated pulmonary metabolism of carcinogens. *American Journal of Respiratory Cell and Molecular Biology* 44:583–590.

40 Anisimov, V.N. 2003. The relationship between aging and carcinogenesis: a critical review. *Critical Reviews in Oncology/Hematology* 45:277–304.

41 Birnbaum, L.S. 1980. Altered hepatic drug metabolism in senescent mice. *Experimental Gerontology* 15:259–267. Birnbaum, L.S., M.B. Baird. 1978. Induction of hepatic mixed function oxidase in senescent rodents. *Experimental Gerontology* 13:299–303. Baird, M.B., L.S. Birnbaum. 1979. Increased production of mutagenic metabolites of carcinogens by tissues from senescent rodents. *Cancer Research* 39:4752–4755. Birnbaum, L.S., M.B. Baird. 1979. Senescent changes in rodent hepatic epoxide metabolism. *Chemico-Biological Interaction* 26:245–256.

42 Birnbaum, L.S. 1991. Pharmacokinetic basis of age-related changes in sensitivity to toxicants. *Annual Review of Pharmacology* 31:101–128.

extremely noxious disease controlled by water purification, and this control was and is a salubrious outcome of epidemiological studies.[43]

Cancer is a complex disease that results from the interaction of host susceptibility and environmental conditions. Molecular epidemiology is focused on environmental factors as they affect host DNA, and it is these factors which offer a fertile field for cancer prevention. Activated chemical carcinogens from the environment form covalent adducts with host DNA (and cell proteins), as indicated throughout this book. While the specific organ that is at risk is where the carcinogen/DNA adducts work their havoc resulting in cancer, surrogate cells such as blood cells, exfoliated buccal cells and placenta are also found to contain the specific DNA adducts. It is the DNA adducts in non-target cells (i.e., surrogate cells) that become one kind of biomarker for cancer.

The measurement of adducts in surrogate cells offers the potential for cancer prevention through early recognition of populations at risk (i.e., prior to the appearance of overt cancer), identification of individuals at risk, the detection of pre-clinical disease and the evaluation of carcinogen protection measures. Molecular epidemiology is a relatively new area of research and its full potential for cancer prevention is yet to be attained[44]—when it is attained, it will be one more significant area where the Miller's research has both enabled and preceded related research with a potentially great impact on cancer prevention.

Laws Regulating Carcinogens

It is a simple matter to access extensive information concerning international treaties as they relate to carcinogenesis. One website is titled: "*International Treaties and Initiatives: Chemicals and Wastes: Multinational Environmental Initiatives*" which makes it manifestly clear that there is international concern for carcinogens in the environment. Almost all scientists share this concern, as

43 Vinten-Johansen, P., H. Brody, N. Paneth, S. Rachman, M. Rip. 2003. *Cholera, Chloroform, and the Science of Medicine: A Life of John Snow*. New York: Oxford University Press.

44 Wogan, G.N. 1992. Molecular epidemiology in cancer risk assessment and prevention. Recent progress and avenues for future research. *Environmental Health Perspectives* 98:167–178. Perera, F., I.B. Weinstein. 2000. Molecular epidemiology: recent advances and future directions. *Carcinogenesis* 21:517–524. Perera, F.P., J.B. Herbstman. 2008. Emerging technology in molecular epidemiology. What epidemiologists need to know. *Epidemiology* 19:350–352. Yuspa, S.H., P.G. Shields. 2008. Etiology of cancer: chemical factors. pp. 203–209, In: V.T. DeVita, T.S. Lawerence, S.A. Rosenberg (Eds.) *Cancer: Principles & Practice of Oncology*, 8th Edition, Philadelphia: Lippincott Williams & Wilkins.

did Elizabeth C. and James Miller. I believe that the Millers metabolic studies of chemical carcinogens and other areas of research that they gave rise to are of unquestioned scientific merit. While the Millers work contributed to a general concern for chemicals in the environment, they were among a large population of scientists who gave rise to this concern. With regard to metabolism of chemical carcinogens, the Millers were without peer.

Teratogenesis and Drug Metabolism are Related to Metabolic Activation

The subject matter of this section is not specifically mentioned in the Millers' Biographical Memoir of the National Academy of Sciences (See epigraph at beginning of Chapter 3). Obviously, it could have been and for that reason, is included herewith.

Teratogenesis refers to the induction of developmental defects which occur in both animals and humans. While developmental defects may occur due to a number of toxic effects on the developing organism, here the defects referred to are those induced by chemicals in the human embryo.[45] The known chemical teratogens comprise a small group, but the list will, in all likelihood, be much larger as more research is done. Historically, chemical teratogens were recognized by relating anatomical abnormalities of the newborn to exposure to a particular toxic chemical. Non-anatomical defects in fetal development may be manifested as changes in mental competence, behavior, metabolism, nervous coordination, and many other conditions—these defects are often missed at birth.

It has only been relatively recently that metabolism of potential teratogens has been associated with teratogenesis. Metabolism of teratogens is complex and may be mediated through the cytochrome P450 enzymes of the liver. And as indicated previously, the discovery of liver microsome enzymes occurred in Elizabeth and James Millers' laboratory.

A description of metabolic activation of chemical teratogens bears a remarkable similarity to that of the metabolic activation of chemical carcinogens: "... [S]tudies based primarily on rodent models suggest that many potential embryotoxic xenobiotics are actually proteratogens that must be bioactivated by enzymes such as the cytochrome P450 ... to teratogenic reactive intermediary metabolites." The highly reactive, electron deficient metabolites

45 Juchau, M.R. 1989. Bioactivation in chemical teratogenesis. *Annual Review of Pharmacology and Toxicology* 29:165–187.

bind covalently (irreversibly) to electron-rich cellular macromolecules DNA (and RNA and proteins) of the embryo, forming covalent-bonded adducts, thus irreversibly altering gene activity with abnormal development ensuing.[46] The study of chemical carcinogenesis has led to new insight in the much feared realm of abnormal fetal development. And, as chemical carcinogenesis studies have the potential of reducing human cancer, so too do studies of teratogens have the potential of reducing anomalies in human development.

Most people dutifully take a prescribed drug and think only that the particular drug will improve or cure the medical problem. They rarely think that the drug is, in fact, a xenobiotic chemical and that like other xenobiotics, many drugs are broken down by cytochrome P450 enzymes.

The liver is the principal site for drug metabolism. Phase I metabolism, mediated by Cytochrome P450, primarily results in oxidation of the drugs. Phase II results in biotransformation of the drug, by conjugation with a host substance, resulting in its greater solubility and ease of excretion. Drug metabolism was considered earlier in Chapter 3 in the section "The paradox of cancer-causing enzymes in normal humans."

It is abundantly clear that an understanding of drug metabolism is not possible without a consideration of the liver enzymes first isolated in the Miller laboratory.[47] Elizabeth and James Miller were not pharmacologists, but, as is obvious in this brief description, it is essentially impossible to understand drug metabolism without reference to the enzymes first reported in the Miller laboratory.

---

46 Wells, P.G., L.M. Winn. 1996. Biochemical toxicology of chemical teratogens. *Critical Reviews of Biochemistry and Molecular Biology* 31:1–40. See also: Juchau, M.R. 1989. Bioactivation in chemical teratogenesis. *Annual Review of Pharmacology and Toxicology* 29:165–187. O'Brien, P.J., B.F. Hales, P.D. Josephy, A. Castonguay, J. Yamazoe, P.F. Guengerich. 1996. Chemical carcinogenesis, mutagenesis, and teratogenesis. *Canadian Journal of Physiology and Pharmacology* 74:565–571.

47 Conney, A.H. 2003. Induction of drug-metabolizing enzymes: A path to the discovery of multiple cytochromes P450. *Annual Review of Pharmacology and Toxicology* 43:1–30. Lynch, T., A. Price. 2007. The effect of cytochrome P450 metabolism on drug response, interactions, and adverse effects. American Family Physician 76:391–396.

CHAPTER 5

The Family Origins of Elizabeth Cavert Miller: New York, Ireland and Scotland

"We have spent an exciting 35 years together in trying to ferret out some of the properties of chemical carcinogens and the changes they cause in their target tissue. When we started our work together in 1942, chemical carcinogenesis was a rather limited field that had attracted only a small number of investigators." The studies have "given much information relevant to the prevention of cancer." These quotations were taken from the 1977 Presidential Address to the Denver meeting of the American Association for Cancer Research, Inc. by Elizabeth Cavert Miller.[1] She alerted her audience that she and her husband, James Alexander Miller, were pioneers in how chemicals cause cancer and she concluded that those studies became relevant for the prevention of that dread disease. I would term these thoughts as the *alpha* and *omega* of the Millers' research. The *alpha*, the beginning, concerns how certain chemicals begin the transformation process that leads normal cells to become malignant cells. The *omega*, the significance of their results is the prevention of cancer.

Elizabeth's collaborative research with James Miller led, in part, to my desire to write not only about their seminal investigations, but also about their lives. "Brilliant scientist", "superb cancer researcher" and "fantastic scholar" are phrases used to describe Elizabeth. Those phrases apply no less to James. However, in this chapter and the following, I concentrate on Elizabeth. I suppose that Elizabeth comes first for several reasons: She grew up only a few blocks from where I live in the Saint Anthony Park neighborhood of Saint Paul, Minnesota; I enjoyed lecturing about her work during the many years that I taught a course in the biology of cancer; and she graduated from the University of Minnesota where I have been a faculty member since 1970. How could I not consider Elizabeth first? The life and times of James and his family, of equal importance, are described in a later chapter.

I believe that family history and life experience play a major role in providing insight into brilliant accomplishment. Further, just as many readers wish to learn about the family and personal details of the lives of notable public figures, so too is there a desire among some, and that includes me, to know something

1 Miller, E.C. 1978. Some current perspectives on chemical carcinogenesis in humans and experimental animals: Presidential address. *Cancer Research* 38:1479–1496.

 | DOI 10.1163/9789004286801_006

of the personal life of a woman who spent countless hours at the laboratory bench digging into the complex and arcane molecular reactions that lead to cancer. Here then is a glimpse of Elizabeth Cavert Miller's family over several generations. It may hold a clue as to the why of her accomplishments.

Elizabeth's family was far from ordinary. Generation after generation, through profound effort coupled with strong moral values and innate ability, her predecessors achieved success in varying endeavors. One of her ancestors was a pioneering founder of a beautiful village still flourishing two centuries later, while other ancestors were successful farmers; still others were gifted preachers and distinguished professors. This was the family tradition of achievement in which Elizabeth matured.

Family history and life experiences, including upbringing, education and socioeconomic condition, may be considered nurture. Elizabeth was a child of 9 at the onset of the Great Depression, and certainly the economically difficult times that began even before she was a teenager affected not only her but also her parents and siblings and where and how they lived. Exceptionally few, and this did *not* include the Cavert family, escaped the economic disaster of the depression. Elizabeth's birth in Minneapolis, Minnesota and her early years in Saint Paul and then Anoka, Minnesota are the "nurture" of her early life and that nurture, added to family history, may have contributed to who she became.

All of the nurture in the world, however, will not yield a brilliant scholar without an innately exceptional nature. And nature here refers to what I commonly refer to as "intellect". Like many of her family members, Elizabeth enjoyed an abundance of intellect. It was probably this combination of family tradition of excellence, Elizabeth's early years in Minnesota and an acute intellect that enhanced her growth and ambition and gave the scientific world the remarkable Elizabeth Cavert.

This family history includes an account of Elizabeth's New Jersey and New York ancestors, as well as bit of her Irish and Scottish antecedents. Elizabeth's pioneer ancestors moved to Charlton, New York, from New Jersey and so the focus of her history begins with Charlton.

Charlton, New York

Elizabeth Cavert's father and many of her other ancestors were born in Charlton, New York, while her mother was born in nearby West Charlton. Charlton is by no means a city and even now it appears to be little more than a hamlet, albeit a spread-out hamlet of 32 square miles. Charlton had a population of less than 100 with about 20 families when Elizabeth's father and his siblings were growing up. Recently the town boasted nearly 1,500 families with a population of 4,133 (census of 2010)—it may *appear* to be a hamlet but sometimes appearances deceive. The village was formed in 1792 and, because many of its settlers came from Freehold, New Jersey, Charlton was originally named New Freehold.

Charlton, New York is located in southwest Saratoga County between Schenectady to the south and Saratoga Springs to the northeast with only 40 miles, as the crow flies, from the state of Vermont. The rivers and the gently rolling hills make for a beautiful landscape that one might compare to parts of Scotland.

If a traveler to Charlton happens upon the Sweetman Cemetery which aptly shares its name with Sweetman Road (County Route 55) north of the Charlton Road (County Route 51), one will notice a number of prominent names in this old burying place with the first interment occurring in 1778. These are the founders of Charlton, and many of them are Elizabeth Cavert Miller's antecedents. There are no less than 44 graves with Cavert carved in the stone. Both Franklin Delano Roosevelt and John Fitzgerald Kennedy described the United States as a nation of immigrants—and Elizabeth Cavert's family was not an exception. Her Cavert immigrants had their origin in Northern Ireland while another family branch, the McCreas, hailed from Scotland.

An Early Cavert

John Cavert (Elizabeth's great, great, great grandfather) was born in 1734. There is little known about the locale of his birth other than that it was Northern Ireland. John came to the land that became the United States before he was 30 years of age. He settled near Freehold, New Jersey, not far from the Atlantic coast. John met and married Catherine Sweetman (1742–1789) of Princeton, New Jersey, which is located less than 25 miles from Freehold. The wedding occurred in 1762 at the Old Tennent Church of Freehold. Catherine was most likely from Freehold originally, because both her father and mother are buried in the Old Tennent Church cemetery.

With family and friends John settled in New Jersey. Soon John sought to find a site that would provide a more prosperous and fulfilling life than that experienced in Freehold. John Cavert was dispatched by his neighbors to nearby New York state. He travelled the Hudson River north to Albany, then overland to Schenectady followed by a trek into the unsullied but propitious wilderness. John sought land in the old Indian (Mohawk branch of the Iroquois) hunting ground that became a portion of the Kayaderosseras Patent.[2] John judged this land to be a good town site and so too did many of his fellow townspeople who subsequently followed John the following spring.

The reason for starting this family history with John Cavert is that he was a significant ancestor of Elizabeth Cavert and he gave rise to a number of other significant antecedents. The baptismal records of John and his wife's three children are recorded in the Old Tennent Church of Freehold, New Jersey. These records are of interest because the spelling of the family name is not that which Elizabeth knew as a young person. The three children and their baptismal dates in parentheses were William Calvert (14 May 1775), Michael Calvert (10 May 1778) and Mary Calvert (20 June 1779).[3] The question at this point is what happened to the "l" in the family surname? No "Calvert" appears on any headstone in the Sweetman Cemetery in contrast to the 44 headstones with the surname engraved as "Cavert". In a letter to me (2012) the Reverend Barbara McDonald, pastor of the Old Tennent Presbyterian Church, speculated that the minister at the time the Calvert children were baptized, may have spelled the family name the way *he* thought it should be spelled. Perhaps upon relocating in New York, the family reverted to the way *they* thought it should be spelled.

In a letter dated 12 February 1915, Kate Underhill of Gloversville, New York, refers to ancestors from the north of Ireland who, in the late 1700s, spelled the name Calvert. Underhill also professed to have wondered why the "l" was dropped, but the why was "more than I know." No shame is associated with the "l" containing name. Quite the contrary, there seems to be a bit of pride that the Cavert family might be related to Sir John Calvert, the Lord Baltimore. John Calvert was knighted by James I of England, who was also James VI of Scotland (this is the James of the *King James English Bible*). Sir John, as he was known subsequent to his knighthood, was the father of Cecilius Calvert, 2nd Lord Baltimore (1605–1675). The Calverts were intimately associated with the founding of Maryland. I would like to interject the possibility that a genealogist with computer skills might enjoy finding the connection, if it exists, between the Calverts of Maryland and the Caverts of New York and Minnesota.

Several generations and many years after John Cavert founded Charlton, Walter I. Cavert (1854–1936) was born in Charlton. Walter married Elizabeth Brann in 1886. Elizabeth Brann is notable, among other reasons, because her

2 The Kayaderosseras Patent, signed by Queen Anne of England, was the largest land grant made during colonial times. The 406,404 acres (635 square miles or 1,645 square kilometers) contained most of what became Saratoga County, New York.

3 Symmes, F.R. 1904. *History of the Old Tennent Church*, 2nd Edition., Cranbury, N.J.: George W. Burroughs. page 220.

mother's maiden name was McCrea,[4] which links the family history to a notable Scottish highland castle (see next section). Walter and Elizabeth were the grandparents of Elizabeth Cavert Miller. The grandparents had four children: William Lane Cavert (1887–1977), father of Elizabeth, Samuel McCrea Cavert (1888–1976), Walter Dudley Cavert (1891–1986) and Inez May Cavert (1893–1958). We will explore more of that generation later.

Elizabeth's Heritage Includes a Castle in Scotland

The McCreas of Elizabeth Cavert's family can be traced directly to Scotland and to a William McCrea who was born there in 1688 some five generations prior to Elizabeth. Why mention the McCreas? I mention the McCrea family because of their restored clan home which is located in Wester Ross[5] in the North West Highlands. The castle is situated on a tiny island at the point where the waters of Loch Duich to the south and Loch Long to the northeast join Loch Alsh to flow west to the sea. The castle, not far from the Kyle of Lochalsh and the bridge to the Isle of Skye, is known as Eilean Donan, was probably (perhaps?) built as protection from the marauding Vikings.

Eilean Donan, originally constructed in 1220, was in ruins for many years. In 1511, the MacRaes became the Hereditary Constables of the castle.[6] Early in the 20th century, Lieutenant Colonel John Macrae-Gilstrap and Farquar Macrae began a two decade endeavor to restore the castle to its original elegance with eminently successful refurbishment and rehabilitation (Figure 5.1). The Macrae cemetery, a mile or two southeast of Eilean Donan, holds many Macraes, long departed ancestors of Elizabeth.

Curiously, Elizabeth's link to Eilean Donan is not only through the McCreas on the Cavert's side of the family, but also on her mother Mary Mead's side of the family. The castle is named Eilean Donan after the Irish religious hermit Saint Donan (b.?, d. 617 or 618) who was sent to the island about 580 AD by Saint Columba (b. Ireland, 521, d. Iona, 597). It was Columba's and Donan's desire to convert the local barbaric heathen (the "pagan Picts"?) to Christianity.[7] It is

4 The American family name McCrea is a variant spelling of the Scottish Macrae, with or without a capital letter "R" as in MacRae.

5 "Wester Ross" is the western area of Ross and Cromarty of Scotland.

6 Historically, a constable was an individual charged as the keeper or defender of a castle.

7 Saint Columba, while an early Christian missionary to Scotland, was not the first. Saint Ninian of Whithorn preceded Columba. Whithorm is located about 62 miles from Dumfries in the south of Scotland.

FIGURE 5.1 *Eileen Donan Castle, located in the North West Highlands of Scotland.* PHOTOGRAPH BY THE AUTHOR.

believed that the Scottish family of Donnan can be traced to the religious communities founded by the saint. How do the Meads, and thus Elizabeth Cavert, connect to the Donnans?

Alexander Donnan (1759–1843) was born in southwest Scotland. He travelled to Saratoga County, New York in 1777 and is buried in the West Charlton cemetery. Several generations later, Mary Elizabeth Mead, a descendent of Alexander Donnan, married William Lane Cavert and became the mother of the Elizabeth Cavert Miller. She was also, of course, the mother of Elizabeth's siblings, Jane McCrea Cavert and Henry Mead Cavert, discussed later.

The Donnan connection to Eilean Donan is perhaps a bit more tenuous than the McCrea relationship. Saint Donan[8] came to Scotland about 580 AD. I strongly believe that no genealogical program will ever connect the religious hermit of the castle and the island which bears his name to Elizabeth's ancestor, Alexander Donnan. But indeed she is descended from Scottish Donnans; who is to say that they are not connected in some way with the good Saint? Add that possible connection to the more probable McCrea connection and

8 Variant spellings are the rule and not the exception, so little note should be made of "Donan" vs. "Donnan". *Both* spellings are used in Edwin Sprott Towill's 1996 *Saints of Scotland*, 3rd Edition, Edinburgh: Saint Andrews Press.

one can see a reason for family pride and just possibly a desire to augment or extend the achievements of the past.

Elizabeth Cavert's Father

William Lane Cavert was the eldest of the four children of Walter I. Cavert and his wife Elizabeth Brann. William grew up on the family farm located north of Charlton Road on Sweetman Road. William wrote a fascinating account of farming in upstate New York in the early 1900s.[9] The 93-acre farm of William's youth was relatively small by today's standards but it gave young William a superb exposure to diversified farm management. The Caverts ordinarily had only eight cows. Early on, the milk was transported to a cheese factory. Later the milk sold at 80 cents per 10-gallon can. About 30 acres were planted in hay making it a major cash crop. The average harvest of hay annually was about 10 tons, which sold at about $12 per ton. Harvesting of the hay required about 3 weeks of intense human- and horse-power. Elizabeth's grandfather added to the family income by dealing in apples. The Cavert farm produced about 100 barrels of apples selling for as little as 40 cents to as much as $1.50 per barrel.[10] Inferior apples used to make cider sold for as little as 15 cents per barrel. Dressed hogs were going for about 8 cents per pound. The farm income also came from the sale of milk, eggs, chickens, rye, pears and sundry other farm products—clearly this was diversified farming which worked well for the Cavert family at the time. Records were kept and estimates of farm income were later made by Elizabeth's father who ultimately made a career in teaching farm management in Minnesota—the beginnings of which he learned well on the family farm on Sweetman Road in Charlton, New York.

The productivity of the land assured the Caverts of food to eat, so in a sense, times were not hard. But in another way, life was difficult or even primitive by modern standards. There was no electricity, no telephone, no radio, no refrigerator, no car, no indoor bathroom or running water—and radio, television, computers and smart phones were years in the future.[11]

---

9 Cavert, William Lane. 1958. *The Farm Community and Farming in Charlton, Saratoga County, N.Y. Circa 1900.* Typescript, 12 pages single spaced.

10 For urban dwellers, it is worthwhile to point out that the U.S. Dry Measure of a barrel is 4 pecks. Each peck comprise 8 quarts. A quart of apples thus went for a little over one cent to a little less than 5 cents. Of course, cider apples went for less than a half cent per quart.

11 Schmidt, W.J. 1978. *Architect of Unity: A Biography of Samuel McCrea Cavert.* New York: Friendship press.

It is a tradition for Presbyterians to hold education in high regard and the Charlton Freehold Presbyterian Church on Charlton Road was no exception. The original church was founded in 1786. The current church, a fine example of Greek Revival architecture, was built about a decade prior to the Civil War. It has colored glass windows memorializing both the Cavert and Brann families. Associated with the church was the Academy building where education of members' children took place. It was erected in 1859 and served for classes until 1913. The Academy building was destroyed by fire in 1946 and never rebuilt.

William Lane Cavert's mother and one of his uncles attended the Academy, as did William and his siblings.[12] The Presbyterian Charlton Academy, run by the Albany Presbytery,[13] had but a single teacher in 1900. Because William was taught to read at home, he did not begin at the Academy until he was nine. While the Academy was small, it did afford Elizabeth's father the opportunity to study algebra, plane geometry, four years of Latin (but no Greek), and four years of English and history (American, English and Ancient). That curriculum would still serve a pre-college student well a century later. Obviously the sole teacher, Mary E. Callaghan[14] of Lake Charlton, New York, who had completed only one academic year, 1881–1882, at Wellesley College in Wellesley, Massachusetts, was talented and multifaceted to an amazing degree. Much of what she presented as a teacher must have reflected the high standards of Wellesley College—and what she learned by extensive reading elsewhere. Notwithstanding, Elizabeth's father, by his own admission, showed the least ambition of the four Cavert children in acquiring an education. He characterized his attitude toward education as a "lack of interest".[15] Despite this "lack of interest", his attendance for only the senior year at Schenectady High School after the academy was adequate for admission to nearby Union College in Schenectady. He began his studies at Union in 1906 and graduated in 1910.

William Cavert, Elizabeth's father, was encouraged by his father to study law. However, it was agriculture that caught his fancy. The failure to study law

---

12 Much of the information of the Cavert and Mead families was obtained from a family document: Mitchell, Jane Cavert, *Cavert and Mead Ancestors*, typescripgt dated 4 March 1995.

13 Presbyteries are district governing bodies of the Presbyterian Church (USA).

14 Jane A. Callahan, Archivist, Wellesley College, Wellesley, Massachusetts confirmed Mary E. Callagham's attendance at the college.

15 Perhaps there was a bit of exaggeration in the claim of "lack of interest" for higher education. Note that William Lane Cavert graduated from Union College and Cornell University with undergraduate degrees. These were followed by a Master's degree from the University of Minnesota and finally the Ph.D. from Cornell University. A rhetorical question: Did William Cavert truly exhibit a "lack of interest" in education? I think not.

may have been a disappointment to his father, but it did not prevent the father from taking young William to a farmer's Institute in Saratoga Springs, located northeast of Schenectady, where he met Professor Herbert H. Whetzel[16] of the Cornell Agricultural College. With encouragement from Whetzel, William enrolled in Cornell University at Ithaca, New York. Soon he became interested in agricultural economics and farm management with George F. Warren[17] as a mentor. The distinguished Warren was not only a professor at Cornell but also an economics adviser to Franklin D. Roosevelt during the time Roosevelt was governor of New York and later while he was president of the United States. Cavert obtained his B.S. degree in farm management—a relatively new academic discipline with the benefit of perhaps the finest agricultural economics professor of the day.

After graduation in 1912, Elizabeth's father (Figure 5.2) received a job offer from the University of Minnesota. Will, as he was sometimes known, was pleased indeed to accept the position in the University's Agricultural Extension Division as an instructor in farm management. Agricultural programs at the University of Minnesota had been around for some time having started a year after passage of the Hatch Act of 1887.[18] Cavert took advantage of the opportunity to take University courses and earned his master's degree in Agricultural Economics in 1914 while working full time. Subsequently, he took sabbatical leave (with his family) from May 9, 1928 until June 30, 1929 for graduate study in agricultural economics at Cornell University, where he received an early Ph.D. (1929) in this burgeoning academic field.[19]

Cavert's original instructor's salary at Minnesota was $600 for six months beginning October 1st, 1912.[20] The appointment was extended to a full year

16 Whetzel, Herbert Hice, 1902–1944, Professor of Plant Pathology, Cornell University. See: Whetzel, H.H. 1918. *History of Phytopathology*, Philadelphia, W.B. Saunders, Co.

17 Warren, George F., 1874–1938. See *Time* Magazine cover story of November 27, 1933 and biography by Stanton, B.F. *George F. Warren, Farm Economist*. Ithaca, New York: Cornell University Press, 2007.

18 The Hatch Act of 1887 was sponsored by Missouri Congressman William Henry Hatch. Federal dollars were provided to aid states, including Minnesota, in establishing agricultural experiment stations. There is another "Hatch Act", 1939 amended 1940, sponsored by Senator Carl Hatch of New Mexico. The more recent Hatch Act concerns corrupt practices in federal elections. It has absolutely nothing to do with agricultural research.

19 One source stated that William Lane Cavert's Ph.D. was granted in 1928, others stated it was 1929. The actual date was 25 September 1929 according to the Office of the University Registrar of Cornell University, Ithaca, New York.

20 Salary data as well as certain other information about William Lane Cavert were obtained from the Archives, University of Minnesota. Elsewhere, Elizabeth Cavert Miller reported

FIGURE 5.2 *William Lane Cavert, father of Elizabeth Cavert Miller. Date of image unknown.*
PHOTOGRAPH COURTESY OF THE MILLER FAMILY.

beginning in 1914 with the salary of $1200. Later in 1914, Cavert became a Farm Efficiency Agent at $1800 with $1500 of that sum paid by the United States government. After 1914, over 80% of Cavert's salary was provided by the federal government.

---

that all faculty at the University of Minnesota, including her father, took a one-third cut in pay during the depression. All faculty took the cut so that no faculty would be terminated. ECM did not specify what year of the depression that this occurred.

FIGURE 5.3 *Former home of the Cavert family, located in Saint Anthony Park of Saint Paul, Minnesota.*
PHOTOGRAPH BY THE AUTHOR.

William returned to New York to marry Mary Elizabeth Mead in 1914, after which the newly married couple settled in their home in the Saint Anthony Park area of Saint Paul, Minnesota (Figure 5.3).

By 1934, Cavert had been recruited and hired to become the chief statistician, and subsequently, Director of Research in Farm Management, for District Seven of the Farm Credit Administration[21] in Saint Paul.[22] The Farm Credit Administration was a Franklin D. Roosevelt "New Deal" agency established in 1933. The agency was designed to assist farmers to refinance mortgages at a lower interest rate with a longer time for repayment than would be possible with regular bank loans during the extraordinarily difficult years of the Great Depression. William was now earning $6,000 per annum, five times the sum at the first full year of his career, and this amount was described by his son Mead as seeming like a "mint". Cavert retired from the Farm Credit Administration in 1957.

21 District Seven of the Farm Credit Administration was comprised of four states, *viz.*, Michigan, Minnesota, Wisconsin and North Dakota.

22 The McArdle Laboratory for Cancer Research Memorial Resolution for Elizabeth Cavert Miller, dated 1 February 1988, lists Cavert's position as "Director of Research in Agricultural Economics, Federal Land Bank, *Minneapolis* (Italics added)".

A now deceased emeritus professor in the Department of Agricultural and Applied Economics, University of Minnesota, Philip M. Raup, spoke of his and his colleagues close friendship and professional esteem for Will Cavert and his family many times to this author. It is evident then, that while Cavert was working for the federal government, he maintained close contact with university colleagues and friends.

Elizabeth's family heritage included the siblings of her father. Samuel McCrea Cavert (1888–1976) and Walter Dudley Cavert (1891–1986), were elected to Phi Beta Kappa at Union College. Both studied theology. The brothers had an unmarried sister, Inez May Cavert (1893–1958) who graduated from Vassar College, Poughkeepsie, New York, in 1916.

Brother Samuel, known to his friends as "Mac", was for many years the executive leader of the Federal Council of Churches of Christ in America, founded in 1908.[23] He, more than anyone else, led the Federal Council to merge in 1950 with the National Council of the Churches of Christ in the United States of America (NCC). Samuel became the first general secretary of the NCC, serving from 1950 to 1954. Subsequently, he worked to establish the World Council of Churches (WCC), for which he served as Executive Secretary of the New York office.

Dudley Cavert, like Samuel, was a clergyman of significance. After graduation from Union College, BA, 1907, he received his Bachelor of Divinity degree, 1915, from Union Theological Seminary in the City of New York. Subsequently, he received the honorary Doctor of Divinity degree from his alma mater Union College. Walter served as minister at churches in several New York cities. He then became the Field Director of the Board of Christian Education in New York. He was the Vice Moderator of the United Presbyterian Church, USA, in 1961.

Inez May Cavert, the youngest of the four children of Walter I. Cavert and Elizabeth Brann Cavert, never married. She worked as an administrative associate[24] at the headquarters of the National Council of Churches in New York City. She, as did many other Caverts, published scholarly papers.[25]

---

23 Schmidt, W.J. 1978. *Architect of Unity*, New York: Friendship Press, is an excellent account of the distinguished clergyman Samuel McCrea Cavert.

24 Inez May Cavert worked on the research staffs of the Federal Council and National Council of Churches and was an editorial associate of Information Service published by both councils. See obituary, *The New York Times*, 29 November 1958.

25 Example: Cavert, I.M. 1953. Status of women in the local church. Problem or Opportunity? *Pastoral Psychology* 4:21–27. Her title, as author of this paper, is: "Research Associate, National Council of the Churches of Christ in America". *The New York Times* of 24 November 1940 carried an article about her earlier research with the title of "Ask equal voices in the churches."

Elizabeth Cavert's Mother

Early in his career at the University of Minnesota, William Lane Cavert married Mary Elizabeth Mead (1890–1976) at the Mead family home. Mary Elizabeth, was the daughter of Henry Winne Mead (1838–1911) and Mary Elizabeth Smeallie (1850–1927).[26]

Mary Elizabeth Mead was born on 8 September 1890 in West Charlton, New York. She died in Minnesota on 21 February 1976. Her Mead ancestors were pioneers who settled in Morris County, New Jersey located in the northern part of the state west of New York City. The Meads arrived in Morris County in 1695. Almost a century later in 1786, a descendant of the early Meads, John Mead and his wife, moved about 150 miles north to Charlton by means of an oxen-drawn wagon. They took up farming in a new location. Their grandson, Henry Winne Mead and Mary Elizabeth Smeallie, had the daughter Mary Elizabeth Mead who became the wife of William Lane Cavert.

The Mead family home was next door and south of West Charlton's United Presbyterian Church. The church, located 6 miles northwest of Charlton, was built only a decade prior to Mary's birth. It was and is a fine wood frame Gothic structure similar today as it was when Mary Elizabeth was born. Mary Elizabeth and William Lane Cavert travelled to Minnesota after they were married.

Mary Elizabeth Mead was, as her husband to be, well educated. She attended a private academy in Schenectady, New York and then matriculated at Vassar College, Poughkeepsie, New York, where she received her bachelor's degree in 1912, the same year that William obtained his undergraduate degree from Cornell. Following Vassar she was a graduate student in home economics at the Teachers College of Columbia University in New York City for two years until she married. The marriage of Mary and William resulted in the birth of three children; Jane McCrea Cavert (b. 1918), Elizabeth Cavert (b. 1920, a primary subject of this book), and Henry Mead Cavert (b. 1922). The siblings of Elizabeth are described in the next chapter.

As Mary Cavert was college educated including graduate work, it is not surprising that she enjoyed reading and was a regular patron during her marriage at the local Saint Anthony Park Carnegie Library in Saint Paul only a few blocks from her home. Elizabeth's mother, as most mothers of the time, helped and guided her children in their school homework. Her son Mead, however, said that his very bright sister Elizabeth did not need much help—if any at all. The observation that Elizabeth did not need much help was an early sign of the

26 Unlike many of Elizabeth Cavert's other relatives, the Meads, Henry Winne and Mary Elizabeth (Smeallie), are buried in the cemetery of West Charlton's United Presbyterian Church.

intelligent, self-directed, persistent and highly motivated scholar that she was to become.

Although Mary Calvert stayed at home to mind the children, she was no "wall flower" and was described by her son Mead as a "social person". This social person, college graduate, family homemaker, and loving mother must have passionately desired for the success of her children—and this included Elizabeth of course.

William and Mary Cavert came from traditional Scottish Presbyterian Churches in upstate New York. Elizabeth's father described the local Charlton Freehold Presbyterian Church as fundamentalist. He went on to state that most in his community considered the teaching of evolution dangerous and that Charles Darwin was thought to be a bad influence on the world. However, by the turn of the century times were changing in small town New York—or at least in Charlton. Elizabeth's father told of a Charlton Presbyterian minister, who came to believe that the fundamentalist beliefs of the time were not intellectually tenable. While such liberal views may have been taught, traditional Presbyterian cultural values of hard work, commitment to education, abstinence from liquor, intellectual curiosity and duty to church and family and community tenaciously persisted. The values persisted not only in the Charlton community but also with William Lane Cavert and his children. It is not surprising that two of William's brothers, uncles of Elizabeth and her siblings, became distinguished Presbyterian clergymen as described above.

This chapter has chronicled some of Elizabeth Cavert's family history. It should be noted that Elizabeth's ancestors included keepers of a castle in Scotland, founders and land-holders of a New York community, and her farmer father who became a professor and government administrator. Her uncles were prominent churchmen. Her mother graduated from a fine liberal arts college and went on to graduate school. Elizabeth's parents nurtured their children with a love for education and the children responded to this love. Elizabeth's sister married a professor and her brother is a distinguished emeritus medical school professor and administrator. Earlier in this chapter, the concept of nurture was considered and how it may have affected Elizabeth's life. If nurture includes the impact of family for guidance and stimulation in learning, and a rich family history, then Elizabeth certainly enjoyed an incredibly high level of familial support for excellence.[27]

---

27 As noted in the text, both of Elizabeth's parents were college graduates. This was not common in 1920, the year of Elizabeth's birth.

CHAPTER 6

More about Elizabeth

Elizabeth as a Child Lived in Saint Paul, Minnesota

Saint Anthony Park, a neighborhood in the northwest part of the city of Saint Paul, is friendly and quiet and near the geographic center of a metropolitan area well over three and a half million people. You would never suspect the size of that metropolitan population if you gazed across the large fertile fields of alfalfa, oats, corn, wheat and soybeans of the University of Minnesota's Minnesota Agricultural Experiment Station, adjacent to Saint Anthony Park. "The Park", as Saint Anthony Park is known by its residents, has a small town feel. This is especially true if you walk on the rolling hills of Luther Seminary.

A modest but well used Carnegie Public Library, two nice restaurants, a locally owned neighborhood bookstore by name of "Micawber's" (perhaps named for the Wilkins Micawber in Dicken's *David Copperfield*), a generously stocked neighborhood grocery store complete with a butcher behind the meat counter (how many supermarkets today have on-site butchers?), a full service branch bank, a pottery shop with beautiful hand thrown pots, jars, cups and plates, a post office, a couple of filling stations with mechanics on duty, and other small businesses contribute to the village atmosphere. The streets are winding, and as at the Seminary many are hilly. In fact, Saint Anthony Park's College Park has an immense hill that entices children of all ages to perilously sled downhill on the snow at breath-taking speeds. Neighborhood children abound because Saint Anthony Park has always been attractive to families. Many of those families have an academic association with Luther Seminary[1] or the University of Minnesota's Twin Cities campuses. As might be expected because of the academic residents, the Park's elementary and junior high schools enjoy reputations for quality education. Saint Anthony Park was, and still is, a good place for a child to grow up.

While Charlton, New York, Saint Paul and Anoka (more of which later), Minnesota were of great and continuing significance to her family and antecedents, Elizabeth was actually born in the small Hill Crest Surgical Hospital, in Minneapolis, Minnesota.[2] Hill Crest operated from about 1910 for the next 25 years or so and was located at the southwest corner of Franklin Avenue West

1 Luther Northwestern Theological Seminary became Luther Seminary in 1994.

2 White, S.M. 1930. Minneapolis as a Medical Center. *Annals of Internal Medicine* 3:774–791.

 | DOI 10.1163/9789004286801_007

and Harriet Avenue, not far from downtown Minneapolis. The hospital had 75 beds and a school of nursing. After the hospital closed, the building was converted into an apartment house which is still open and running.

Elizabeth was born May 2, 1920.[3] Her first seven years were spent in a modest Saint Anthony Park home at 1443 Grantham Street, Saint Paul, Minnesota (Fig. 5–3), which has changed little in the past nine decades. She moved with her family to Anoka, Minnesota in 1929. Elizabeth graduated from the Anoka high school in 1937, and then moved back to Saint Anthony Park for her undergraduate studies at the University of Minnesota. Let us first meet her siblings and then consider her early years in Saint Anthony Park.

Elizabeth's Siblings

William and Mary Mead Cavert already had one daughter, Jane McCrea Cavert (1918–2008), when on May 2, 1920, a second daughter Elizabeth Cavert was born. The daughters were followed by a son, Henry Mead Cavert, born in 1922 (Figure 6.1). Elizabeth was known to remark that the three children of William and Mary Cavert were brought up with the idea that they would all be good students and they would all go to college. The parental expectations were fulfilled—plus a good deal more.[4]

Elizabeth's older sister, Jane McCrea Cavert, was a 1940 graduate in dietetics at the University of Minnesota. She interned at Western Reserve University Hospitals in Cleveland, Ohio. After the internship, she married John Edwards Mitchell[5] in the parental Caverts' Saint Paul home in February of 1942. Elizabeth served as maid of honor. Jane took a job at the Richards Treat Cafeteria[6] in Minneapolis. With a university degree in dietetics, she managed the gift shop, served as hostess, and finally came to manage the kitchen.

---

3 Elizabeth died October 14, 1987. Her final illness is discussed in Chapter 8.

4 The reader may wish to refer to the last paragraph of the preceding chapter

5 Much of the information on Professor John E. Mitchell was gleaned from Faculty Document 2077 dated 3 November 2008, University of Wisconsin, Madison "Memorial Resolution of the Faculty of the University of Wisconsin-Madison on the Death of Professor Emeritus John E. Mitchell.

6 Richards Treat Cafeteria and Food Shop operated from 1924 until 1957 at which time it was razed to make room for the new First National Bank of Minneapolis. The cafeteria was opened and managed by Lenore Richards and Nola Treat, both of whom were formerly on the faculty of home economics at the University of Minnesota. A historical account of this Minneapolis landmark written by Kate Roberts, *Richards Treat Cafeteria 1924–1957*, is found in the November 2007 issue of *Minnesota History*, pp. 282–295.

FIGURE 6.1 *The Cavert children in 1926. From left to right are Elizabeth, Henry Mead, and Jane McCrea.* PHOTOGRAPH COURTESY OF THE MILLER FAMILY.

John Edwards Mitchell, Jane's husband, was born in San Francisco in 1917. He graduated in agricultural biochemistry from the University of Minnesota in 1939. He lived with his parents very close to the Saint Anthony Park family home of his future wife. A reason for the proximity of homes is that John's father worked at the Lake States Forest Experiment Station[7] on the campus of the University of Minnesota. John, who was better known throughout his life as Jack, married his neighbor Jane in February of 1942; he began graduate

7 What was known as the "Lake States Experiment Station" underwent mergers and reorganizations to become what is now known as the "United States Department of Agriculture Forest Service, Northern Research Station" which is located, as was its parent antecedent, on the Saint Paul campus of the University of Minnesota.

studies in horticulture and plant pathology at Louisiana State University, Baton Rouge, Louisiana. Only a miniscule few escaped major changes in their life during World War II and Jack was no exception. He served a total of 42 months, including duty in the Pacific Theater of Operations, as a Lieutenant, U.S. Navy. After World War II, Mitchell earned his Ph.D. from the University of Wisconsin-Madison. He and Jane returned to Madison in 1956 where John was appointed an assistant professor of Plant Pathology and then rose through the academic ranks, eventually becoming professor and chair of the department (1975 to 1980).

Early in their married life, the Mitchells shared a home with Elizabeth and James Miller in Madison, Wisconsin. John and Jane Mitchell became the parents of three sons: William, Charles and Robert. John E. Mitchell and his wife Jane Cavert Mitchell died in 2008.

Elizabeth's brother, Henry Mead Cavert,[8] attended Anoka High School in Anoka, Minnesota. With his father's encouragement and his sister's example, he majored in agricultural biochemistry at the University of Minnesota on the Saint Paul Campus. Mead's academic adviser at the University of Minnesota was William Martin Sandstrom,[9] the same faculty adviser as Elizabeth's. Henry Mead Cavert graduated with an undergraduate degree from the University of Minnesota in December 1942.

I asked Mead Cavert, who later earned the M.D. degree, and subsequently became the Associate Dean of the University of Minnesota Medical School, if he had been a "pre-med" during his undergraduate days.[10] Dr. Cavert expressed his strong view that education prior to entering medical school should be in the liberal arts and/or in the sciences and education as a "pre-med" was an inappropriately narrow academic background. The University of Minnesota was one of the pioneering institutions which sought well-rounded students for admission to the study of medicine.

The United States entered World War II in December of 1941, and thus the war was well underway at the time of Mead's graduation from undergraduate

8 Henry Mead Cavert, who has been known by his middle name since birth, was extraordinarily helpful to me in the early stages of writing this book and some of his comments are included in this chapter.

9 Sandstrom, William Martin was born in Devils Lake, North Dakota in 1897. He obtained his Ph.D. from Professor Ross A. Gortner in 1927. He was an Associate Professor of Agricultural Biochemistry at the time he advised Elizabeth Cavert and Mead Cavert.

10 The question was raised in part because *agricultural* biochemistry (author's italics) perhaps seems a bit atypical as a background for the study of medicine.

studies. Mead's father took him to the railroad station when he left Saint Paul for cadet training as a meteorologist in the United States Army Air Corps.[11] Mead emerged from the Army Air Corps in January 1946 with the rank of Captain. He lost no time in marrying his sweetheart June Lorraine Sederstrom of rural Litchfield, Minnesota when he returned from the war and military service.

In January 1946 Mead was admitted to the graduate program in physiology at the University of Minnesota. Because of the encouragement of two of his professors, Drs. Maurice B. Visscher and Nathan Lifson, he found that he was working for both the M.D. and Ph.D. degrees. After he received his M.D. (1951) and Ph.D. (1952),[12] he was recruited to the Department of Physiology, where he taught and performed research. Mead Cavert served the medical school for many years as associate dean.

Emeritus Professor of Pediatrics at the University of Minnesota, Dr. C.C. Clawson, told me that, while deans of the medical school came and went, it was Mead Cavert who kept the medical school on an even keel. The director emeritus and now deceased founder of the University of Minnesota Comprehensive Cancer Center, Dr. John H. Kersey, revealed that it was Dr. Cavert who found funds for Kersey's international study in Scandinavia while he was a medical student. Kersey was en route to Sweden in 1963 for a medical student elective rotation when he became stranded in New York City because of an airline ticket mix-up. John had neither money to return to Minnesota nor to get a new ticket for Sweden. A telephone call to Dr. Cavert, whom John Kersey referred to as "the medical students' friend", resulted in an immediate loan that permitted continued travel to Stockholm. John told this author that Mead "definitely saved the trip—a truly helpful dean." These and many other distinguished physicians and scientists remember Mead and June with much affection and respect. Dr. Cavert retired from the university in 1992.

11 The military branch of the armed forces that deals primarily with aviation has undergone several name changes as well as administrative organization. The United States Army Air Corps was established in June of 1926. It was discontinued as an administrative organization in March of 1942. However, the United States Air Corps continued as a combat branch of the United States Army until the National Security Act of 1947 when it was incorporated into the United States Air Force.

12 Cavert, Henry Mead, 1952, Ph.D. thesis, University of Minnesota: *Metabolism of Carbon 13-Labeled Substrates and Kinetics of Deuterium Oxide Exchange in the Isolated Heart.*

Early Years

Elizabeth was a child at a time that retrospectively seems extraordinarily peaceful and bucolic—she grew up to become internationally known and extensively honored among cancer scientists. Whether it was her family's care, affection, and role modeling, or her inherited intellect, or her times and environment, or a combination of all of these, that permitted and encouraged her nascent talents to mature into the internationally acclaimed scientist that she became, I suspect we will never know. But that does not prevent us from enjoying the story of a delightful girl growing up in mid-America prior to World War II.

Little is recorded from Elizabeth's early childhood other than the somewhat cryptic comment attributed to her mother, that the small child Elizabeth was "lazy". Since that is in such a profound contrast to her life in high school and thereafter, I must question what "lazy" meant. The mother's comment seems to have arisen because the toddler Elizabeth sat in the middle of the floor with a Sears Roebuck or Montgomery Ward catalog for hours. Those catalogs were enormous during the 1920s, so I would like to inject my version of what was transpiring. Could it be that the little Elizabeth was learning about the multiplicity of material things available in that world outside of her modest home in Saint Anthony Park? Instead of saying that she was "lazy", perhaps it would have been more appropriate to say that at a very early age, she had learned to *concentrate, to observe and to learn*. Her skills in *concentration* certainly served her well in later years, and the same can be said about her ability to observe and to learn. Perhaps the relative inactivity associated with concentrating on those giant catalogs led to the mistaken notion that she was "lazy". I simply find it difficult to believe that Elizabeth was lazy. After reading the above about the "lazy" girl, one of Elizabeth's daughters suggested that perhaps her mother was a slow starter or that she was "late to bloom". June Cavert (Elizabeth's sister-in-law) could not believe that Elizabeth was lazy "at any age". Nor can I.

Elizabeth attended the Gutterson Elementary School,[13] in Saint Anthony Park where her older sister Jane was two years ahead of her (Elizabeth and her sister continued their elementary education for one year in Ithaca, New York when the family moved there as her father finished his Ph.D. at Cornell, University). Partly because my children grew up in Saint Anthony Park and

---

13 Gutterson Elementary School was located across the street from the Saint Anthony Park United Methodist Church at the intersection of Hillside and Como Avenues in Saint Paul, Minnesota. It was built in 1887 and demolished in 1953. It was located about one mile from Elizabeth's 1443 Grantham Street home.

because I have no evidence to believe otherwise, I must conclude that Elizabeth Cavert's childhood years in the "Park" were happy. None of the many people with whom I have spoken have indicated otherwise.

Anoka, Minnesota

The Cavert family briefly resided in Ithaca, New York from May 1928 to June 1929[14] while William Lane Cavert finished his Ph.D. at Cornell University. They returned to Saint Anthony Park after Cornell, but only for a brief time as a house swap with a friend resulted in the family moving to Anoka, Minnesota in 1929.[15] Anoka was then a small town located at the confluence of the Rum and Mississippi Rivers. Since it was only about 18 miles northwest, as the crow flies, from the University of Minnesota Saint Paul campus, it was practical for William Lane Cavert to commute by automobile to his job at the University. Elizabeth's brother Mead described Anoka in that era as the sort of small town in which it was likely that you knew everyone you met or saw while walking down the street. Today (2015) it is a suburban city with a population of over 17,000.

The Cavert family years in Anoka (1929–1939) were not the best of economic times, coinciding as they did, with the Great American Depression. The move to Anoka was, at least in part, due to financial reasons since Dad Cavert could and did cultivate vegetables and raise poultry to enhance the family income. The elder Cavert had hoped for an increase in pay in 1929 but instead found his salary reduced. It was a time when a family was lucky to have a home or, as many said when I was a boy, to have "a roof over your head".

The years in Anoka were not necessarily a blissful time for Elizabeth's mother, at least at the beginning. The mother lacked household help, and, probably of greater importance, she lost her social contacts in the city. Inasmuch as she had graduated from Vassar College with its comfortable amenities, it may have come as a shock to find that there was no inside toilet in her first Anoka home.

---

14 University of Minnesota Archives date 1928 as the beginning for graduate study at Cornell University with the Ph.D. dated 1929. The University of Wisconsin dates the study at Cornell as one year earlier, see: Miller, Elizabeth and James. Interviewed by Laura Smail. 1985. Compact disc. University Archives and Records Management Services. Madison, Wisconsin.

15 See Miller, J.A. 1993. Elizabeth Cavert Miller (1920–1987). In: *Women in Chemistry and Physics: A Bibliographic Sourcebook.* Grinstein, L.S., R.K. Rose, M.H. Rafailovich (Eds). London, Connecticut, Greenwood Press, pp. 420–433.

That house was located at 1330 Fourth Avenue South a few blocks north of the Mississippi River in what was then the southeast part of Anoka. The Cavert family lived in their Fourth Avenue home for about five years. At that time, it was the only house on the block and had an alfalfa field behind it. There was a "fair-sized" vegetable garden next to the house.

A Mrs. Howell had an acre of land next to the Mississippi River (south of the Cavert home), and she rented that land to the Caverts. It was there that a much larger garden was started with strawberries, watermelons, squash, green beans and other crops.[16] Chickens and turkeys were kept as well. The chickens produced eggs and Elizabeth, when she was age nine and for several years thereafter, and her younger brother, Mead, were kept busy peddling eggs in the neighborhood. But selling eggs was only the beginning of the money-making endeavors for the children. William Cavert's garden needed attention, and the two younger siblings toiled away at 5 cents an hour tending the crops.[17] Beans were grown in long rows, harvested and sold by the pail when they matured. Young Elizabeth helped with the turkeys and chickens too.[18] The turkey work must not have been much fun, because after that early experience Elizabeth fervently disliked the bird for the rest of her life. In addition to sales of agricultural products, the two younger Cavert children sold popular magazines, such as the *Saturday Evening Post*, *Good Housekeeping* and *Colliers*, door to door.

While the younger siblings were toiling in the field or selling magazines and garden produce, the more mature sister Jane faithfully assisted mother Mary Mead Cavert with household tasks—housecleaning, meal preparation, and cookie baking, all without parental remuneration.

---

16 The author remembers with considerable nostalgia eating fresh vegetables directly from our family garden during the Great Depression days. Little did I know that my father planted his large garden more for economic reasons than for the pleasure of eating fresh and tasty veggies. Despite economic necessity, I savored the delight of garden-fresh corn and beans (and okra and tomatoes)—super market vegetables simply did *not* and still *do not* compare!

17 The nickel an hour may seem penurious by today's standards but it should be noted that the father joined the University of Minnesota at $600 per year and while he earned significantly more with his Ph.D., his income was still only $1800 per year during the late 1920s: Salary data courtesy of the University of Minnesota Archives.

18 The University of Minnesota for a number of years has had an active turkey research program now housed in the Department of Animal Sciences. Dr. Benjamin Pomeroy was hired by the University in 1934 and he soon became well known in turkey and chicken diseases. However, it would seem that the Cavert endeavor for turkey growing preceded Dr. Pomeroy's tenure at the University by a number of years and was due solely to William Lane Cavert's initiative.

In contrast to baby Elizabeth who earlier may have been thought to be "lazy" by her mother, the child Elizabeth was described by brother Mead as "always on the go, a hard worker, industrious, and enthusiastically receptive of her dad's projects". This attitude seems to have brought pleasure to William Lane Cavert who much appreciated the hard work of his two younger offspring. Elizabeth assimilated the elements of economics (and thrift) from her father who was, after all, a Ph.D. in agricultural economics. The father specialized in the *business* of farming—Elizabeth and Mead were learning entrepreneurial skills from him at an early age. While Elizabeth *learned* those skills, it is perhaps noteworthy that when she had daughters of her own, she was sympathetic with their lack of excitement in selling Girl Scout Cookies. Despite learning to sell door-to-door quite successfully as a child, Elizabeth Cavert never pushed her daughters to "peddle" those very popular cookies or anything else.

Obviously the young Elizabeth performed hard manual labor for her nickel an hour. Thus, might this hard-working girl be thought of as a drudge? Hard working yes, but definitely not a young girl slogging away at thankless chores. Although not many girls would have persevered with such physical work during hot and humid summer days, and some of her personal endurance and strength may have overlapped those of boys, and she feared not to get her clothes dirty and rumpled, she was *not* a drudge. Neither was she, as brother Mead said, a "frilly silly" girl. It would seem simply that she was more interested than perhaps many other young people in work that included gardening for her father, selling garden produce, and peddling eggs and magazines door-to-door. Elizabeth was an active, hard working girl and perhaps this foreshadowed her incredible role as a hard working cancer researcher in the years to come.

Later the Caverts moved west across the Rum River to their new location at 2415 North Ferry Street in the northwest part of the small town of Anoka. Ferry Street runs parallel to the Rum River. Elizabeth and Mead had to go back across the river to attend Washington Elementary School located on 6th Avenue North. The school was torn down in 1964 but a school by the same name at the same site exists there today. Unfortunately, highway development took precedence over history and the Cavert house on North Ferry Street, and a number of other homes as well, were demolished.

Elizabeth attended both Anoka Junior High School and Anoka High School. She graduated as valedictorian of the Anoka High School class with 81 students on June 10th, 1937 (Figure 6.2). Obviously, as the number one student in her high school class, she must have been a candidate for membership in the National Honor Society. Indeed she was and she served the organization as Vice President. She was President of the science club *Der Wissenschaft Verein*

FIGURE 6.2 *Elizabeth Cavert graduated first in her class from Anoka High School in 1937. Photographer unknown.*
PHOTOGRAPH COURTESY OF THE MILLER FAMILY.

and her brother Mead was a chair of the chemistry division during the time of her tenure in office. Elizabeth worked for the school newspaper, the *Anokahi*, and played the cornet in the high school band. Her brother and sister were also in the band and they both played clarinet.[19] Her brother Mead judged that the three siblings, with regard to playing in the band, were "mediocre plus".

Mead described Elizabeth as the "star student in the family". She was "very bright and very well organized and self-disciplined". Evidence to support her brother's assessment is found in her high school academic record. During her junior year, for example, she earned "A's" in English, Solid Geometry, American History, Chemistry,[20] and Trigonometry. In those days, "Deportment" was graded and here again "A" appeared on the record.

Summers at Fort Plain, New York

Minnesota ordinarily has simply elegant summers, providing the mosquitoes do not become too aggressive and the summer heat wave(s) with high humidity are short and few in number. So, one might ask if Elizabeth enjoyed her summers in Anoka, Minnesota. This question refers, of course, to the years that followed her gardening and sales endeavors. The answer to the rhetorical question is that she alternated between summering in Fort Plain, New York, and Springfield, South Dakota. Summers in Springfield will be discussed later. Concerning the New York summers, as noted previously, Elizabeth's mother Mary Mead Cavert was a Vassar College graduate. Mary had a lifetime friend from her time at Vassar College, Alice Weller, who had a large home in Fort Plain, New York. The 2000 census population of Fort Plain was a little over 2,000. Fort Plain is in Montgomery County, New York, on the Mohawk River and Interstate

19 Information about Elizabeth Cavert's High School experiences was obtained through the courtesy of Executive Director Todd Mahon, Anoka County Historical Society, Anoka, Minnesota.

20 Elizabeth had a physics and chemistry teacher by name of James Isaac Watters (1908–1989). Watters was a superb teacher who encouraged students and ran a science club. Evidence that he indeed was an outstanding scientist and one well equipped to teach chemistry obtains from the fact that he subsequently earned a Ph.D in chemistry from the University of Minnesota, 1943 (under the direction of Professor I.M. Kolthoff), and later taught at the University of Chicago, the University of Kentucky, and ultimately became chair of analytical chemistry at Ohio State University. Elizabeth was indeed fortunate in having a fine chemistry teacher while yet in high school. It was thought by family members that Mr. Watters should be given credit for stimulating Elizabeth's interest in chemistry.

highway 90 between Utica and Albany. Both sister Jane and Elizabeth, initially with their mother and subsequently without, traveled a number of summers to stay with Alice Weller. What a treat during the nasty part of the depression to escape mosquitoes and the sometime summer heat waves of Minnesota to stay with "Aunt Alice" in cool and comfortable Fort Plain where the maximum mid-summer heat is 80 F (27 C) degrees or lower.

Winters in Anoka, Minnesota

Of course, all Minnesotans know that there is a season completely free of mosquitoes. Winter! And that season brings snow and a time for skiing. William Cavert would outfit all of his children on simple wooden skis and off they would go. Elizabeth loved this outdoor winter activity even when the temperature was minus 20 degrees Fahrenheit! It is recorded that she enjoyed skiing more than her siblings, and it was said, "she ate it up." Thanksgiving day, celebrated on the last Thursday in November, was traditionally the beginning of the ice skating season. Recall that Ferry Street where the Cavert's home was located in Anoka runs parallel to the Rum River. The father and children needed only a short hike to the frozen river to skate. Temperature was no concern; Dad Cavert only wanted the ice to be safe. Elizabeth, being hearty and strong naturally, loved it.

When winter temperature moderated, Dad Cavert encouraged his children to play tennis. A tennis court was only a block or so away. Elizabeth tried hard, but despite being strong, she was not an "athletic type", according to her brother. An important family game was croquet. Both Dad and Mom Cavert had played the game in New York on the farm front lawns where they had grown up. William Lane Cavert became particularly formidable at the game; in fact, it was said by young members of the family that he was a "killer" at croquet and showed no mercy whatsoever when playing with other family members and that included the children. As might be expected, croquet was played at the Fourth Avenue South home in Anoka. Elizabeth, as in tennis, was not spectacular at the game but her brother described her attitude in playing as "determined", which it must have been to even stay in the game with the relentlessly winning, croquet-playing father.

From the preceding vignettes, it becomes obvious that her family was very important during the early years of Elizabeth's life. Mother Cavert played the piano, and Sunday evenings were a time to gather the family. Elizabeth's brother Mead describes Dad Cavert as a person who was tone deaf. Elizabeth may not have been tone deaf, but neither was she an aria singer. In fact, family members recall that she was asked *not* to join the choir at church. To add further insult to injury, she stated that she was asked "not to sing out loud" in church as her voice was "not good". Elizabeth told this to daughter Linda when Linda was rejected from school choir in the 6th grade. It would seem that the previously disparaged mother sought to reduce the pain of her similarly tone-challenged daughter. While Elizabeth's vocalizing left something to be desired, she enjoyed singing with her family the old time favorites, such as Robert Burns' *Coming Through the Rye*, Andrew Lang's *The Bonnie Banks O' Loch Lomond* and Stephen Foster's *Old Folks at Home*, also known as *Way Down Upon the Suwannee River*. Added to those old time favorites were hymns from the *Methodist Hymnal*. Anyone as old as the author of this biography will know how precious are memories of family togetherness. The ubiquitous television has all but obliterated family togetherness—Elizabeth and her siblings were blessed in growing up at a time prior to eyes being glued to the small screen.

One need not have an exceptional voice to enjoy music and as an adult Elizabeth attended concerts, and enjoyed, to some extent, listening to music at home. However, for much of the time Elizabeth seemed to prefer silence. This is in contrast to her husband James who indeed enjoyed listening to classical music on the home stereo.

Elizabeth was an unsurpassed scientist in her adult years. Were there early indications of what she would become during the Anoka years? Mentioned previously were her outstanding grades in school. Add to that her brother's description of her as a "book worm". She treasured the book that she received at Christmas each year from her father's sister Inez May Cavert. She read the classics authors, such as Charles Dickens and Sir Walter Scott, as did her brother and sister. In this way, Elizabeth and her siblings were following the examples of their parents, William and Mary Mead Cavert, who were avid readers. But there are many avid readers who do not become world-renowned scientists, as Elizabeth did.

Springfield, South Dakota

Family trips were a learning experience and a lot of fun for the Cavert children. Elizabeth's father had an interesting uncle by the name of Samuel McCrea Brann (1863–1944). The uncle, granduncle to the Cavert children and called "Uncle Mac" by Elizabeth and her siblings, settled in Springfield, South Dakota.[21] Samuel was multitalented and was perhaps one of many family role models for Elizabeth. Successively, he was a graduate of Princeton University, Princeton, New Jersey, a cattle rancher in Montana, the owner of a meat market

21 Springfield is located in southeast South Dakota in Bon Homme County just across the Missouri River from Nebraska. The 2010 census reported a population of 1,989.

followed by a grocery store in Iowa and a hardware business, then a clerk in his father-in-law's bank. He was self taught in many respects; he learned Arabic and French as an adult so that he could read classic works in their original language. He eventually however became interested in farming and the business of buying and selling farm land.

Family visits to Springfield were exciting for the children. Elizabeth's sister Jane remembered going there in the 1920's by train. Mother Cavert would gather her three children, Jane, Elizabeth, and Mead, and head for the railroad station to visit Samuel McCrea Brann and his wife Jane. It was an overnight ride on the train to Yankton in southeast South Dakota. Springfield was only a few miles further west but only accessible by a local "milk train". The Mom and children traveled in a manner that the author of this book always yearned to do—in the caboose! Later they traveled to Springfield in the comfortable and speedy Whippet automobile[22] of Dad Cavert.

Uncle Mac's farm was actually located in Nebraska across the Missouri River from Springfield in fertile bottom-lands. The farm, cultivated by Niobrara Indian families, eventually became flooded with the waters of the Lewis and Clark Lake, formed by the Missouri River Gavin Point Dam started in 1952 by the United States Army Corps of Engineers.

While Uncle Mac and his wife had no children, they had the things that keep kids happy. The children enjoyed a swing and a sandbox which occupied their long summer days. More fun was the old horse, described as a "slow nag", which they rode around town. It may have been a slow nag, but Elizabeth and her brother and sister considered this old horse a true prize-winner. Rides on the nag were safe for the children because Dad Cavert or Uncle Mac led the horse. There were, of course, treats for the children at the Springfield home with its huge vegetable garden. Luscious strawberries were in abundance as were all the other good things to make for bountiful meals.

Did the Cavert children have any notion at all of the economic and social horrors of the great depression when they had the great good fortune of summer visits to Springfield, South Dakota and Fort Plain, New York? I doubt it.

Elizabeth Cavert's Father Taught His Children about Agriculture

During automobile trips to South Dakota and camping excursions to northern Minnesota, Will Cavert enjoyed teaching his three young ones about crops, fences, farm machinery, soil and its moisture content and livestock. He taught

22 The Whippet automobile, discontinued in 1931, was made by the Willys Corporation which became famous for manufacturing military Jeeps during World War II.

his children and later, his grandchildren, how to distinguish a prosperous farm from a poorly managed farm. He taught them to discriminate between a high-yielding corn crop and an agricultural disaster. He even exited the automobile to examine crops and prognosticate how they would fare. On a related note, Cavert took his children to the Minnesota State Fair, located just next door to the University of Minnesota Saint Paul Campus, an easy walking distance from the Cavert's Saint Anthony Park home. Probably his prudent Presbyterian upbringing led him to the notion that the Midway entertainment was a "waste of money", and because money was in fact scarce during the depression years, the Midway with its rides and sideshows was strictly off limits to the children. Instead, they were taken to the cattle barn where they learned how to identify the various breeds of cattle found in Minnesota at that time. Surely, those were fulfilling and fun learning experiences from a father who wanted his children to know something of the land and its people who support us all. While Dad Cavert was an agricultural economist with a Ph.D., his son-in-law John E. Mitchell described him as "a farmer at heart".

Return to Saint Paul and the University of Minnesota

Even during her high school years, Elizabeth was showing considerable interest in science. Anoka High School had a fine science teacher named James Watters who introduced Elizabeth to chemistry and physics.[23] By the time she finished Anoka High School in 1937, she knew she wanted to be a chemist. Initially Elizabeth believed that she should become a chemical engineer, but at the time professed not to know what a chemical engineer did. Her father thought home economics would be fine. Her mother had studied home economics at Columbia University. Her father, being a practical man, believed that home economics could be academically stimulating and good for finding a teaching job, and if Elizabeth was to marry, home economics would be useful there too.

It was commonly believed that women would do better in home economics (also known at that time as "domestic science") where jobs were available than with a major such as biochemistry where there was a dearth of positions. The author's own sister who seemed to have a natural talent for calculus and the higher mathematics was successfully pressured to major in home economics at a state supported college. Many years later, I can only think "what a waste of a superb mind".

Family friend and mentor-to-be, Professor Ross Aiken Gortner, discouraged Elizabeth from her first choice of chemical engineering, for, as he said, it was

23 See footnote 20 this chapter.

"not a very likely profession" for a woman.[24] Gortner suggested that she would do far better in biochemistry. Once Elizabeth heeded the advice of the close family friend and chose biochemistry,[25] she never veered from that path.

While Elizabeth's mother, a Vassar graduate, probably would have liked her daughters to attend her alma mater, Elizabeth instead attended the University of Minnesota. The University of Minnesota choice stemmed from economic considerations (Elizabeth already had a sister at the University of Minnesota and a brother coming along in a few years), and besides, especially during the depression, Minnesota was considered "close, cheap, and good." It remains so well over a half-century later, but I would aver the description to read "close, manageable in cost (but not "cheap"), and outstanding." As an aside, Elizabeth did indeed pay tuition but the $60 per year cost at the land-grant institution, was minimal, or as she said "cheap".[26] For some reason, the practice of tuition remission for faculty children commonly found at many other universities was never observed at the University of Minnesota.

Elizabeth took only science and mathematics classes with the exception of a year of German, four courses in Agricultural Economics, and a course in Public Speaking. I was unable to find any course in history, art, music, English, literature, philosophy or any other liberal arts. Her daughters said that she had hoped to take "non-science" courses at the University when she retired. She earned an undergraduate degree in "Science Specialization" as her transcript and the University Commencement stated, and science was indeed what she studied.

Elizabeth's primary focus in her adult years was science and nature. She certainly attended concerts but this was in deference to Jim's love of music. Similarly when she discovered that a daughter enjoyed art and history museums, Elizabeth went with that daughter and the family to art and history museums. She was a caring wife and mother but her focus in life, science and nature, remained unchanged.

---

24 Miller, Elizabeth and James. Interviewed by Laura Smail. 1985. Compact disc. University Archives and Records Management Services, Madison, Wisconsin.

25 Actually, Elizabeth Cavert majored in Agricultural Biochemistry. The Department of Agricultural Biochemistry was founded in 1913 in the University's Institute of Agriculture. The departmental name was changed to the Department of Biochemistry in 1963 and the department moved to the College of Biological Sciences in 1965 (many years after Cavert graduated from the University). Curiously until relatively recently, the University of Minnesota had two departments of biochemistry, the other being in the Medical School. As of the time of this writing, both departments have merged into a huge organization known as the Department of Biochemistry, Molecular Biology & Biophysics. Elizabeth Cavert minored in economics.

26 "Cheap" is relative. The $60 in 1937 would be equal to about $1,000 in 2015.

FIGURE 6.3
Ross Aiken Gortner was one of Elizabeth's professors at the University of Minnesota.
PHOTOGRAPH COURTESY OF THE UNIVERSITY OF MINNESOTA ARCHIVES.

Two of Elizabeth's Professors

Elizabeth most likely was a lone woman student in a world of male science students. She said that she recalled *no* women on the Saint Paul Campus in any field of study other than home economics.[27] How was it that Elizabeth was admitted to this all male academic enterprise? It probably was helpful to have a father who was a professor. Similarly, it was advantageous that Ross Aiken Gortner (Figure 6.3), Chief of the Agricultural Biochemistry Department, was a friend of the family and was advising Elizabeth about her proposed academic career. However, family members, speaking years later, believe that it was her outstanding high school grades and academic credentials that qualified her for admission. From the perspective of many years after she was admitted to study biochemistry and with my four decades of teaching experience, I side with the family. It was easier then than it is now to turn down an applicant—especially

27 Miller, Elizabeth and James. Interviewed by Laura Smail. 1985. Compact disc. University Archives and Records Management Services, Madison, Wisconsin.

a female applicant. Fortunately for cancer research, Elizabeth was judged to be an outstanding applicant, which she was, and she was *not* denied entrance to the study of biochemistry. It was here that she had the good fortune of not only having Gortner as a family friend but also as a mentor.

Gortner was born on March 20th, 1885 in Holt County, near O'Neill in north central Nebraska.[28] He studied at Nebraska Wesleyan University where he was awarded his undergraduate degree in 1907. He received his master's degree from the University of Toronto the next year and his Ph.D. from Columbia University in 1909. He arrived in Minnesota in 1914, the same year that Elizabeth's mother and father arrived. Gortner guided the research of 87 graduate students, published over 300 papers in scientific journals[29] and was the author of several notable books in biochemistry.[30] He was a fellow of the American Association for the Advancement of Science and of the National Academy of Sciences. He served as chair of the American Chemical Society, president of the American Society of Naturalists and of the national scientific society Sigma Xi. Gortner was on the editorial boards of *The Journal of the American Chemical Society* and the *Journal of Physical Chemistry*. I would judge that this stellar professor served as a role model for Elizabeth. In my own experience, I found that professors who did research and published their results intensified my desire to be a scientist and to go forth and do likewise. I project from my experience that Gortner enhanced Elizabeth's undergraduate experience and augmented her desire to have a career in research.

Of course, Gortner was not the only Minnesota professor that influenced Elizabeth Cavert. She became an undergraduate research assistant, performing chemical analyses for W. Martin Sandstrom. Sandstrom, a native of Devils Lake, North Dakota, was a bachelor professor who seldom left the lab or classroom. Similar to Gortner, Sandstrom, most likely knew Elizabeth's family as

---

28 Lind, S.C. 1944. Biographical memoir of Ross Aiken Gortner, 1885–1942. Washington, D.C.: *National Academy of Sciences*, 23:149–180.

29 Sandstrom, W.M. 1942. Ross Aiken Gortner. *Register Phi Lambda Upsilon* 27:97–98.

30 I first learned of Gortner through my biochemistry professor, Simon H. Wender (Ph.D. 1938, University of Minnesota), who was one of his many students. In the biochemistry class I took, Wynder used "Gortner, Ross Aiken, 3rd edition, edited by Ross Aiken Gortner, Jr. and Willis Alway Gortner. 1949. *Outlines of Biochemistry*. New York: John Wiley & Sons. It was thick in size and had a green binding. I loved that book and never tired of reading it. Elizabeth's husband to be, James Alexander Miller, when he was an undergraduate at the University of Pittsburgh, used a Gortner textbook which he described as a "very good" biochemistry text. Perhaps Elizabeth and James enjoyed the earlier 2nd edition, published in 1938, as much as I did the 1949 edition edited and updated by his sons.

Sandstrom's home was only one block from the Cavert home. Sandstrom was a co-author[31] of the most widely used lab manual for biochemical procedures and adviser to almost all undergraduate major students in the department.

Sandstrom strongly encouraged his undergraduate assistant and advisee, Elizabeth Cavert, to continue her biochemical studies in graduate school. She did. Good students know that the undergraduate record is the "coin of the realm" with regard to admission into graduate school. Perhaps with this in mind, Elizabeth assiduously tended to her studies and was awarded the Caleb Dorr[32] scholarship and medal for being the top woman scholar (of about 500 female students) on the Saint Paul agricultural campus *each year* of her undergraduate career. Further, she earned the highest grade point average of both men and women in the entire College of Agriculture[33] and was elected to The Phi Beta Kappa Society.[34] As earlier stated, grades were the "coin of the realm" for admission to graduate school. Clearly, Elizabeth earned the right to compete for admission to any academic institution to which she might apply.

Elizabeth as an Undergraduate in 1937

Elizabeth lived in the College Girls' Dormitory on the Saint Paul Campus during her first two undergraduate years[35] while her parents still lived in Anoka,

31 Morrow, C.A., W.M. Sandstrom. 1935. *Biochemical Laboratory Methods for Students of the Biological Sciences*. New York, Wiley.

32 Named for Caleb D. Dorr (1824–1918) who was an early Minnesota lumber and logging businessman and owner of the steamship *Governor Ramsey*. See: Loehr, R.C. 1943. *Caleb D. Dorr and the early Minnesota lumber industry*. Saint Paul, Minnesota: publisher unknown.

33 Miller, J.A. 1993. Elizabeth Cavert Miller (1920–1987). In: Women in Chemistry and Physics, A Bibliographic Sourcebook. L.S. Grinstein, R.K. Rose, M.H. Rafailovich (eds.), Westport, CT: Greenwood Press, pp. 420–433.

34 The Phi Beta Kappa Society, founded in 1776, stands for and promotes excellence in the liberal arts and science. It is the oldest honorary academic society in the United States. It was founded at the College of William and Mary, Williamsburg, Virginia. Mead Cavert, brother of Elizabeth Cavert, believes that Elizabeth was the sole woman to be elected to the society from the Saint Paul Campus of the University of Minnesota in 1941, the year of her graduation.

35 The College Girls' Dormitory, later known as Meredith Hall, was immediately east of where the Student Center is now located. The dormitory was torn down in 1969.

Minnesota.[36] She moved back home with her family to 1443 Grantham Street in Saint Anthony Park of Saint Paul during her junior and senior years. It may have been more economic, and the home was conveniently located only five blocks away from the campus. She did not bother to have her picture published in the University yearbooks known as the *Gopher*. She was making outstanding grades and, whether she consciously chose it or not, grades (and her future career) took priority over yearbooks.

Some of this account (the exceptionally high grades and the apparent lack of time for a yearbook picture) might suggest that Elizabeth was a bit of a drudge while an undergraduate student. *Not true*! A family history relates that Elizabeth kept scrapbooks with notes on dances attended along with scraps from prom dresses and flowers. She saved University of Minnesota football and University theater programs; she made a record of who asked her out and who invited her to dances; and she attended campus talks by notable speakers such as Carl Sandburg. She was a member of the women's home economics sorority, Gamma Omicron Beta, on the Saint Paul campus where social gatherings and parties were numerous. Elizabeth loved those convivial occasions.

Smoking at that time was a new habit for women. About a third of women smoked during the 1940s which is nearly double that of 2015. It was for many an outward manifestation of women's newly discovered independence and sophistication. Smoking could also result from peer pressure. Elizabeth smoked. Her sister-in-law June S. Cavert described her as quite "sophisticated". Certainly smoking was considered sophisticated and perhaps this is why Elizabeth acquired the habit.

Few undergraduates escape romances during their important undergraduate years and Elizabeth was no exception. Elizabeth was extraordinarily attractive, and it would thus have been difficult for male students to fail to notice her. Soon she was engaged to another biochemistry student. This was prior to graduate school. The young couple dated for several years. He was a heavy smoker. Could it be that Elizabeth's smoking resulted from the tobacco habit of her boyfriend? It is possible but, of course, no one knows.

Elizabeth Cavert was an undergraduate at Minnesota from the fall of 1937 until she graduated with a Bachelor of Science degree with "High Distinction" at the University's Sixty-Ninth Commencement June 14th, 1941. Madison, Wisconsin was soon to be her academic domicile.

---

36 The family home at 1443 Grantham Street in Saint Paul was not sold despite the fact the family lived in Anoka from 1929 until 1939. Elizabeth chose not to commute during her first two years at college and thus lived in the dormitory from 1937 until 1939.

Graduate School, Biochemistry and Home Economics

Elizabeth Cavert applied to six graduate schools for admission and financial support. Funds for graduate education was perhaps more important then than it is now because the Great Depression, with its limited dollars available for scholarships, was only beginning to wind down. Elizabeth's exceptional academic record at Minnesota (recall that she had the highest grade point average of the entire Saint Paul Campus of the University of Minnesota and was elected to the Phi Beta Kappa Society at graduation) merited her offers of support from three fine graduate schools; *viz*, Columbia University in New York City, the University of Iowa, Iowa City and the University of Wisconsin, Madison. The University of Wisconsin offered a Wisconsin Alumni Research Foundation (WARF) Fellowship.[37] The WARF Fellowship in 1941, amounting to $700 per annum (almost $12,000 in 2015 dollars), was described by Dr. Henry Pitot, Emeritus Director of the McArdle Laboratory for Cancer Research, University of Wisconsin-Madison, as being "extremely competitive."

Elizabeth judged the (WARF) fellowship to be "much better" than the other offers. Madison was only 266 miles from her home in Saint Paul, Minnesota. She chose Wisconsin. That choice led to finding and falling in love with James Alexander Miller, marriage, graduate work leading to the Ph.D. degree, a career in cancer research and two very special daughters. Elizabeth remained in Madison and never moved again—it was where she obtained her professional education, found her husband and lifetime collaborator, had a home, reared her family, and eventually died. The University of Wisconsin-Madison got all of that because they awarded a well-deserved WARF scholarship to Elizabeth.

---

37 The Wisconsin Alumni Research Foundation (WARF) is a private, not for profit, fund raising organization founded by Professor of Biochemistry Harry Steenbock (1886–1967). Patents obtained by Steenbock and other faculty members have earned $750,000,000 for the University of Wisconsin Madison. The earnings are returned to the faculty and students of the University to support research. Steenbock developed ultraviolet irradiation of food—especially milk—that enhanced vitamin D content. The bone disease rickets was virtually eliminated because of the increased availability of vitamin D. Increased exposure to the sun also increases vitamin D in the skin. Increased vitamin D *may* decrease vulnerability to several forms of cancer—certainly Steenbock and the Millers would have been pleased if this proves to be true (for example, see: Moan, J., A.C. Porojnicu, A. Dahlback, R.B. Setlow. 2008. Addressing the health benefits and risk, involving vitamin D on skin cancer, of increased sun exposure. *Proceedings of the National Academy of Sciences* (USA) 105:668–673).

Elizabeth was off to Madison and the University of Wisconsin in the autumn of 1941. Once again the issue of home economics arose as it had at Minnesota. Recall that Elizabeth's father wanted her to major in home economics during her undergraduate studies. Her father reasoned that it was an academic area in which she could be productive and, in addition, it would be useful in her home if she married. While this was a practical view, Elizabeth felt that she had an inadequate academic background for home economics considering her agricultural biochemistry degree. Wisconsin, however, as many graduate schools of the time (just before the United States' entry into World War II), was having difficulty in placing its male Ph.D. students in suitable positions. Further, the Biochemistry Department acknowledged a sex bias with the assertion that it could find no jobs for women biochemists. Because of this perceived dearth of available faculty positions for women, the Department of Biochemistry would not accept Elizabeth for graduate study in biochemistry alone. They insisted on a joint program, primarily in home economics, but including biochemistry. Perhaps with the combined major, there was at least a fighting chance for a *woman* to get a faculty position in the fiercely competitive market of the early 1940s. A job in home economics, or home economics/biochemistry combined, would be better than no job at all—at least that is probably what the university bureaucrats believed. While she was able to study biochemistry without home economics at the clearly more progressive[38] University of Minnesota, it appears that the administrative powers at Madison won, and Elizabeth took Home Economics courses along with her courses in Biochemistry.[39]

A note concerning the "why" of Elizabeth's eschewing home economics is appropriate here. Elizabeth's daughters believe that her desire not to study home economics had nothing to do with frustrations with the course work nor did they believe that it related to her being a feminist in the usual sense of the word. She simply knew that she wished to study biochemistry and was focused to that end. Nobel Laureate Howard Temin,[40] a colleague of Elizabeth's at the McArdle Laboratory, stated that Elizabeth would want to be remembered as

38 "More progressive" at least with regard to what was expected of women students at this time.

39 While it is true that she was required to take Home Economics courses, she listed all 3 of her degrees as biochemistry. She never cited home economics on her Curriculum Vitae. Elizabeth was relieved when she could switch completely to biochemistry and quit her home economics program.

40 Howard Temin (1934–1994) shared the Nobel Prize in 1975 with David Baltimore for the discovery of reverse transcriptase.

a *scientist who was a woman* and not as a *woman scientist* (italics added). The sex bias against women in biochemistry that was an initial problem for Cavert ended with the acute demand for biochemists of either sex during and after World War II.[41]

Although Home Economics was not of her choosing, Elizabeth was fortunate indeed in having a fine academic adviser named Catherine Personius in that department.[42] Personius endeavored to work out a program for Elizabeth that included a significant biochemistry component. Elizabeth was soon working on bread and the problem of bread texture and strength. The biochemical portion of the study was done in the laboratory of Professor Marvin Johnson.[43] Johnson's laboratory was large with many graduate students, some of whom thought the biochemistry of bread and the curious but talented woman student who studied it to be "very funny".

The newly arrived and talented graduate student from Minnesota proved to be far more than "very funny"; she was exceedingly outstanding in both biochemistry coursework and laboratory research. By the second semester of her graduate career, she was taking the vitamins and metabolism course offered by Carl Bauman. James Alexander Miller was teaching a portion of Bauman's laboratory, and this is where Elizabeth met James.

Any discourse on Elizabeth Cavert's first year at the University of Wisconsin-Madison must at this time include an introduction to her laboratory instructor. This is because James Alexander Miller played such an extraordinarily

---

41 At Drury University, my undergraduate plant sciences professor, Lora Bond (1917–2000), who received her PhD from the University of Wisconsin, was hired during World War II because of a shortage of male professors. She was retained thereafter until retirement because of her many outstanding qualities not the least of which was teaching.

42 Catherine Personius (1904–1994) was born in Elmira, New York. She received her BS from Elmira College in 1925, her MA from Columbia University in 1928, and her PhD from Cornell University in 1937. She was a faculty member at the University of Wisconsin from 1940 to 1943, and it was at this time that she served as graduate adviser to Elizabeth Cavert. Personius returned to Cornell as head of the Food and Nutrition Department, where she strived successfully to develop courses with an emphasis on the chemical and physical properties of food. Perhaps Personius' interest in science was the reason that Cavert was permitted to work on a biochemical problem while being advised, for a short time, in the Home Economics Department.

43 Marvin Joyce Johnson, 1906–1982, was born in McIntosh, Minnesota. He earned his BS, MS, and PhD degrees and was a Professor and sometime Chairman of Biochemistry at the University of Wisconsin-Madison. He is best known for his study of how penicillin is produced by molds.

important role in all aspects of Elizabeth's life, *viz.* her graduate study, marriage, children and professional career.

The following chapter describes James' birth and his family, his undergraduate studies at the University of Pittsburgh, his meeting Elizabeth, the wedding, and James and Elizabeth's career at the McArdle Laboratory for Cancer Research, University of Wisconsin-Madison.

CHAPTER 7

James Alexander Miller—How He Became the Other Strand of the Miller Double Helix

James Alexander Miller's parents, John Herman Miller, Jr. and Emma Anna Stenger Miller, were both reared in Pittsburgh, Pennsylvania. Although neither of them went beyond the eighth grade in school, their limited education did not in any way minimize their great respect for higher education. It was probably due to the economic times that only one of their six sons graduated from college. John and Emma were married in 1900 in Pittsburgh. The Millers were Methodists and the Stengers were Catholic. James rarely saw his paternal grandparents, probably due to the difficulty of James' grandmother Miller interacting with the Catholic Stenger family.

John Herman Miller held the position of prothonotary (chief clerk) in the courts system of Pennsylvania. Later he worked for the *Pittsburgh Press*.[1] "Black Tuesday", October 29th, 1929, a day that is marked by many historians as the start of the Great Depression, preceded by only a few months the loss of John Miller's job in 1930. Job loss so soon after the great stock market crash of 1929 was clearly catastrophic, but John Miller's loss of employment was further complicated by a persistent and serious illness which was thought to have caused episodes of peculiar behavior. He never returned to the *Pittsburgh Press*. James's father moved in with his son Robert and his daughter-in-law Columbia for a few months, after which he was homeless. Later, with the return of his health, John Miller took a job in Washington, Pennsylvania (about 30 miles southwest of Pittsburgh) and eventually returned to Pittsburgh. James A. Miller never became close to his father who was gone much of the time while he was growing up. However, John Miller showed up, unexpectedly, at the University of Wisconsin-Madison commencement when Jim received his Ph.D. It is thought by the Miller family that this was the first and only time Elizabeth Cavert Miller ever met her father-in-law.

1 The *Pittsburgh Press*, now defunct, was at one time the second largest newspaper in Pennsylvania. Mr. Miller had been head of the circulation department.

 | DOI 10.1163/9789004286801_008

James' Birth and Growing Up

James,[2] (Jim to all who knew him), was born to John and Emma on May 27, 1915[3] in Dormont, Pennsylvania. A suburban community with a 2010 census population of 8,593, Dormont is about five miles southwest of what is now Gateway Center, Pittsburgh. The Miller family lived on Pioneer Avenue near the crossing of that street with West Liberty Avenue (Figure 7.1). The Miller family had a large two-story home, suggesting that the occupants were, if not prosperous, certainly comfortable.

James was the fifth of six sons. While three older brothers completed only one or two years of high school and then sought jobs, James and one older

FIGURE 7.1 *Boyhood home of James A. Miller located in Dormont, Pennsylvania.* PHOTOGRAPH BY THE AUTHOR.

2 Some of the Miller history is from notes taken by Helen Miller Alexander from talks with her father. The notes *The Miller Family* are unpublished and undated. Other information was taken from Miller, Elizabeth and James. Interview by Laura Smail. 1985. Compact disc. University Archives and Records Management Services, Madison, Wisconsin. Roberta A. Williams, a niece of James A. Miller, was generous with her time and memories during a meeting with me in 2008.

3 James Alexander Miller died on December 24, 2000.

brother finished high school. Tragedy struck early and often in young James' life. He lost his oldest brother John M. Miller, to lobar pneumonia in 1924. John was only 24 years old. His mother's sister, Rose Stenger, died of pneumonia in 1926. Rose, known affectionately by the boys as "Aunt Po", was a second mother in James' life. Aunt Po assumed much of the responsibility for rearing the Miller brothers and her loss was almost like losing a mother. The youngest brother, Harry, who was three years the junior of James, was killed in 1926 when he was struck by an automobile while playing behind his house in the alley. The boys' mother Emma died unexpectedly and suddenly at the age of 49 of a cerebral hemorrhage (stroke) in 1929 when young James was only 13. The sadness of four family deaths in a six-year period was exacerbated by James' father losing his job in 1930 and becoming seriously ill. The times were described by James as a "dark period" of his life and had the effect of making James "become and remain an agnostic in religious matters."[4]

Even as a young boy, James loved chemistry, and he relished working with the chemistry sets which were given to him. He eagerly looked forward to reading articles on science in the *Pittsburgh Press*. Early indications of James' great interest in chemistry included "experiments" carried out in the basement at home. Jim built a rocket at an early age and prepared fireworks. There were reports of several mishaps or "accidents" (minor explosions) in the basement but James persisted, and perhaps surprisingly, with his mother's blessings. The minor explosions, however, caused major trouble for James from his father who was not amused by the noise or potential for peril.

One cannot get chemicals at the neighborhood hardware or corner grocery store, so young James *bicycled* from his Pioneer Avenue home in Dormont via the long and dangerous three plus mile trek on West Liberty Avenue, *all steeply downhill*, then through the Liberty Tunnel, and finally over the Liberty Bridge across the Monongahela River to downtown Pittsburgh. The object of these daring bicycle rides was to find a warehouse at 711 Forbes Avenue, four miles from his start. The trips for chemicals occurred between 1925 and 1930 and the company on Forbes Avenue, originally known as Scientific Materials Company,

4 Miller, J.A. 1998. The metabolism of xenobiotics to reactive electrophiles in chemical carcinogenesis and mutagenesis: A collaboration with Elizabeth Cavert Miller and our associates. *Drug Metabolism Reviews* 30:645–674.

later became Fisher Scientific Company.[5] There must have been much sweat and pain associated with James' long uphill return trip.[6]

The bicycle ride to the chemical company was, of course, from the Pioneer Avenue home. James' father seems to have had a wish to move further away from the big city, perhaps to provide a more bucolic lifestyle for his six sons. He spoke of his desire to move to Mount Lebanon, a more suburban community southwest of Dormont. James' mother Emma quickly vetoed the notion—she said she had to be able to see street car tracks "from her porch"—and this she could do from the Pioneer Avenue home. The boys' home was sold by their father at Emma's death in 1929 when Jim was 13.

During those extraordinarily difficult times, James' three surviving brothers endeavored to hold the family together, and they succeeded even in the absence of mother and father.[7] Recall that John Miller lost his job and left home during James' childhood. James, now the baby of the family, graduated from Dormont High School in 1933 with average grades, except for high marks in the sciences. His graduation occurred at the depth of the Great Depression and there were few opportunities for Jim to go to college. This was especially true for anyone with "average" grades. He reported that he had an outstanding high school chemistry teacher named Dickerhoff, who may be compared with Elizabeth's superb high school chemistry teacher James Watters. Dickerhoff gave the young James "free run" of the chemistry laboratory, which was a significant help to James, who yearned to be a chemist. While most of James' grades were average, an early and significant indication of his interest and expertise in

5 Chester G. Fisher founded the Scientific Materials Company in 1902, and the plat map from 1923 shows the company location at 711 Forbes Avenue. That company became Fisher Scientific in 1925, and as late as 1975 Fisher was still located on Forbes. However, by the 1970s Fisher had a half dozen locations in Allegheny County. Duquesne University bought the site at 711 Forbes Avenue. By 2006, Fisher Scientific grew to 30,000 employees and was sold for 10.6 billion dollars. The new merged company, with sales of 9 billion dollars per annum, is known as Thermo Fisher Scientific, Inc. The account of young James Alexander Miller buying chemicals from Fisher, and Fisher selling chemicals to a boy, perhaps reveals as much about Fisher and the difficult economy of the time as it does about James Miller.

6 I thank Roberta Miller Williams and her husband Robert for driving me down West Liberty Avenue from what had been the Miller home to the Liberty Tunnel—even in an automobile the hill is awesome. I can hardly imagine the trip on a bicycle.

7 Biographical resources for the Cavert family are abundant. This is not the case with the Miller family. The four Miller boys did not talk much about family. Their reticence may be due to the difficult economy of the time or due to loss of and illness in family or just simply an innate taciturnity.

science was an article authored by James A. Miller with fellow student Walter P. Ketterer published in 1934.[8]

Jim had great determination to learn, especially in scientific areas, as indicated by his chemistry "experiments" and the low-cost ultramicroscope. While still a young man, he wanted a telescope. With the acquisition of glass and abrasives, grinding and polishing, and much patience, James fashioned a telescope from scratch. One need only to refer to instructions for telescope-making to recognize the effort of this endeavor. The completed telescope revealed his capacity to understand the complexity of the project, to plan and organize work, and to tenaciously persist until the job was done, attributes he absolutely continued to display throughout his scientific career. Surely the telescope gave him much pride in accomplishment. I am certain that as he observed the heavens, he saw far more than Orion with his belt and the bright Sirius—the telescope would have revealed a star-crowded sky which may have caused Jim to wonder about life and what it had in store for him. Perhaps he, during these early years, dreamed of a notable career in cancer research.

The other brothers unselfishly supported James while he was in high school and in his desire to go to college. Since scholarships were not available and his brother's funds were limited, James had to find work in order to finance his education with the intention of delaying college only briefly. He found and obtained several jobs, including working as a timekeeper in the welding department in a Bethlehem Steel fabrication plant on the Ohio River near Pittsburgh. For six months, James kept track of the piecework jobs of each welder so that their paychecks correctly reflected the types and numbers of welding jobs each completed (the workers were paid different amounts according to the types of welds).

---

8 Miller, J.A., W.P. Ketterer. 1934. The ultramicroscope as a student project. *Journal of Chemical Education* 11:570–574. Miller and his coauthor correctly point out in their article that the best light microscope of the day (both then and now) had a limit of resolution of about one-tenth micron. There are many biological preparations comprised of particles with diameters much smaller than this. How can one "observe" particles too small to be resolved with the light microscope? The ultramicroscope was designed to detect such particles by a strong beam of light with a microscope objective at right angles to the beam. There is no image when no submicroscopic particles are present. However, as in colloidal solutions, submicroscopic particles are detected with light-scattering phenomena as diffraction halos. The authors point out that the halos are not true images but optical effects caused by their presence. Miller and Ketterer did <u>not</u> invent the ultramicroscope. Rather they reported that an ultramicroscope, too expensive for all but the most affluent high schools, could be inexpensively put together with ordinary high school science equipment and by high school students (one of whom, and perhaps both, proved to be extraordinarily bright).

University of Pittsburgh

James studied chemistry and took an examination for college credit while he was employed at Bethlehem. The self-study saved his family precious dollars. When his brief career at Bethlehem Steel was terminated, Jim enrolled for six months in the night school at the University of Pittsburgh and studied the mathematics required for a chemistry major. The money to pay for the night classes was carefully garnered from his earnings in the welding shop at Bethlehem Steel.

Another of James' early working experiences was washing windows at the University of Pittsburgh. Fortunately for his future studies, he soon moved up to preparing reagents and filling bottles for the freshman chemistry class taught by Professor Elmer Hjort.[9] This he did with financial support, at the rate of 35 cents per hour from a National Youth Administration (NYA) program.[10] With his NYA money, a loan taken out on an insurance policy and help from his brothers, James was able to enroll in an honors program in 1935 and finally became a full time student in chemistry at the University of Pittsburgh. His grades at Pittsburgh were outstanding.

Jim retained a lasting memory of his brothers' encouragement and financial support. When Elizabeth and Jim received the Bristol-Myers Award of $25,000 in 1979 (the award would have about $37,000 purchasing power in 2015,[11] they generously shared that award with the surviving brothers Charles, Christopher and Robert. The money was significant and adequate for one brother to buy a much needed car. That familial generosity is probably known to few others than the readers of this book.

---

9 Elmer Victor Hjort, Ph.D., University of Pittsburgh, 1923. Hjort (pronounced jort) taught chemistry at the University of Pittsburgh from 1920 until 1926 and shared an office with Charles Glen King.

10 The National Youth Administration (NYA) was a Franklin D. Roosevelt New Deal federal agency that operated from 1935 to 1943. The agency paid high school and college students for work study projects or job training. The NYA gave Jim his first job in chemistry of filling reagent bottles at the University of Pittsburgh. Not only did Jim reap benefits from the New Deal but so too did the father of Elizabeth Cavert who, as will be recalled, was materially assisted by another New Deal program (see Chapter Five). A subsequent President of the United States, Lyndon B. Johnson, worked for the NYA. Wealthy people often called Roosevelt a traitor to his class; people of more limited means still honor his memory. There was good reason for both Jim and Elizabeth Miller to so honor Franklin D. Roosevelt.

11 See Appendix for reproduction of *Cancer Research* cover, volume 39, number 1, 1979 showing the Millers who received the Bristol-Myers Award.

James met Professor of Biochemistry Charles Glen King (1896–1988) during his second year at the University of Pittsburgh. King shared an office with Elmer Hjort, mentioned previously with regard to James' reagent bottle filling job, and this is perhaps how Jim met the illustrious Professor King. There is a parallel here with the academic experience of Elizabeth Cavert. Recall that Elizabeth had the good fortune of having a distinguished professor, Ross Aiken Gortner, as an undergraduate mentor at the University of Minnesota. James was soon working 20 to 25 hours per week in an old army barracks[12] for King. He worked as an animal room man, cleaning cages, feeding and otherwise caring for the guinea pigs, mice and rats and one dog used in experiments by King. Of course the Millers, Jim and Elizabeth, used many experimental animals in their joint career and James' experience working for King surely was not wasted. It might be added however that Jim rarely entered the animal rooms at the McArdle—perhaps his early exposure to animal care enhanced his love of the chemistry lab with it test tubes, retorts and beakers.

During his senior year at Pittsburgh, James took King's biochemistry course, and, as recorded previously for Elizabeth, he used the 1938 edition of the textbook *Outlines of Biochemistry* written by Ross Aiken Gortner. Many years later, James recalled that his interest in carcinogenic hydrocarbons was piqued by reading Gortner's textbook. He never lost this interest in cancer-causing hydrocarbons as indicated by his graduate career and subsequent research endeavors throughout his life at the University of Wisconsin-Madison.

While yet an undergraduate at Pittsburgh and prior to his studies at the University of Wisconsin-Madison, James studied with Charles Glen King, an absolute giant in biochemistry. King worked on the water soluble Vitamin C, a vitamin important in the prevention and treatment of scurvy. We, who live in a time when a daily glass of orange juice, or better yet a fresh orange, is taken for granted, have little notion of the former importance of scurvy. Sailors were particularly vulnerable to scurvy because of their appalling diet at sea. Often a Navy ship lost more men to scurvy than in battle, for untreated scurvy is lethal. The Scottish doctor James Lind (1716–1794), in perhaps the first clinical trial in history, experimented on the diet of sailors and showed that orange and

12 Army barracks on college campuses were *not* there to house troops. The buildings were on campus because they were cheap and available as war surplus. The barracks permitted rapid growth of higher education with the return of veterans at war's end. I started teaching at Tulane University in 1961 and even though the World War II had been over for a number of years, the old barracks were still in use and continued to be so for several more years. It is possible that the barracks that were near the chemistry school in James's case was a remnant of World War I.

lemon juice prevented or cured scurvy. Obviously, something in citrus fruit was important in the prevention and treatment of scurvy. Professor Charles Glen King and the professor's graduate student, W.A. Waugh, isolated and purified that "something" which proved to be vitamin C.[13] King was the first to crystallize *any* vitamin in pure form.

Albert Szent-Györgyi also crystallized vitamin C. Szent-Györgyi received the Nobel Prize—King did not. There emerged a controversy, which continues to this day, over the exclusion of King from the award, which many thought he should have received or at least shared with Szent-Györgyi.[14] Be that as it may, it is abundantly clear that King missed the Nobel Prize by only a hair's breadth. Working for a professor that close to the ultimate science prize surely enhanced the already existing creative spark in young Miller. I speculated in the previous chapter that an eminently successful mentor might have enhanced the creative spark for Elizabeth, and I see no reason why it would not have done so similarly for James. Just as I am convinced that King had a marvelously stimulatory effect on James A. Miller, it is probably true that the reverse occurred. King had the good fortune of living to the age of 91, and he lived long enough to be aware of, and to be abundantly proud of, his former undergraduate animal room man who became preeminently important in the study of chemical carcinogenesis.

James A. Miller at the University of Wisconsin-Madison

James was also in close contact at Pittsburgh with one of King's postdoctoral fellows Max Otto Schultze (1906–1998).[15] Schultze had just received his

13 Waugh, W.A., C.G. King. 1932. The isolation and identification of vitamin C. *Journal of Biological Chemistry* 97:325–331.

14 Albert Szent-Györgyi received the Nobel Prize in 1937 for "his discovery of the biological processes of oxidation with special regard to vitamin C and the fumaric acid catalyses"; thus, recognition with the Nobel Prize was for more than crystallization of the vitamin (Cox, G.J. 1937. Crystallized vitamin C and hexuronic acid. *Science* 86:540–542). In this context, let it not be forgotten that Charles Glen King did indeed have priority over Szent-Györgyi in publication on crystalline vitamin C. A brief account of vitamin C discovery and characterization is contained in: Halver, J.C., N.S. Scrimshaw. 2006. *Charles Glen King, 1896–1988. Biographical Memoirs*, 2006. Volume 88. Washington, D.C.: National Academy of Sciences, 2006.

15 I find it curious how academic connections weave a web of associations. Max Otto Schultze, who encouraged Jim to go to Madison, came to the Department of Biochemistry on the Saint Paul campus of the University of Minnesota in 1946 after his tour of duty

Ph.D. in biochemistry from the University of Wisconsin-Madison and was hired as a research associate and paid from a department store research grant to Professor King. This was before the time of the nearly ubiquitous National Science Foundation and National Institutes of Health research grants.[16] Schultze had a high regard for both James and the University of Wisconsin-Madison. The high regard for James was well placed indeed, since James graduated from the University of Pittsburgh in biochemistry with highest honors in 1939. Max Schultze urged Miller to apply for admission to biochemistry graduate studies at Wisconsin, which he did. Not surprisingly, because of his outstanding grades at Pittsburgh and probably a fine letter from the distinguished Charles Glen King, James was awarded an annual stipend of $700 by the Wisconsin Alumni Research Foundation (WARF), just as Elizabeth would receive only a few years later.[17] Miller subsequently described his relationship with Schultze[18] at this time as "absolutely important".[19] Schultze taught James how to pose critical scientific questions, i.e., how to ask a question (in German "*Die Fragestellung*") that will provide new biochemical information and how to

---

in the United States Army. Schultze's appointment at the Saint Paul campus was in the department where Elizabeth had been an undergraduate. I joined the faculty at Minnesota in 1970 and had an office not far from that of Schultze's. I got to know, but only slightly, the man who encouraged Elizabeth's husband to go to Madison. Max Otto Schultze's life prior to academe was in some ways similar to Jim Miller's. Jim was less than affluent prior to and during college. Life was not easy for Jim. Max was a farm boy from Switzerland with no hope of entering university in his home country due to his modest origins and the class distinctions prevailing in Switzerland. It probably was not easy to leave home to be hired out as a farm worker in northern Illinois, but Schultze did so. It was Schultze's fabulously good luck to have a boss who recognized Max as a bright young man. The boss encouraged and helped Max. Max got his education and subsequently helped Jim Miller. Obviously, neither had had it easy but both excelled.

16 The Research Grants Office (later to become the Research Grants Division of the National Institutes of Health) had its origin in January 1946. By August of 1946 study sections, for the scientific and technical evaluation of research grant applications, had been established.

17 The WARF Fellowship has been described as "extremely competitive". Because Jim Miller and Elizabeth Cavert were *both* recipients of WARF Fellowships, they were described by Dr. Henry Pitot as an "intellectual elite" couple.

18 Schultze worked hard for James A. Miller who was but an undergraduate at the time. This may seem to have been unusual for some but it was not for Schultze. Schultze was described as "the hardest working person I ever met"; see: Potter, V.R. 1986. Years with Elizabeth and James Miller: in appreciation. *Advances in Experimental Medicine and Biology* 197:1–9.

19 Miller, Elizabeth and James. Interviewed by Laura Smail. 1985. Compact disc. University Archives and Records Management Services, Madison, Wisconsin.

construct appropriate controls; Schultze stressed the absolute need for rigor in scientific research. Schultze was indeed important for Jim.

James, however, had some reservations about Madison, Wisconsin. The city and university were over 600 miles by highway from Pittsburgh. Chris, James' older brother and a father figure, encouraged him to go and when James agreed, he drove James west to the University of Wisconsin-Madison. In September of 1939 Miller arrived with his WARF fellowship in hand and eager to continue his professional education. He, as Elizabeth, never left Madison.

At the University of Wisconsin-Madison, James was soon studying nutrition and cancer-causing chemicals in the research group of Carl August Baumann (1906–1999) in the Department of Biochemistry of the School of Agriculture (Figure 7.2). Dr. Baumann was then a young assistant professor of nutritional biochemistry. He was collaborating with Professor Harold Rusch of the University of Wisconsin Medical School, who is remembered as the founder of the McArdle Laboratory for Cancer Research in 1940. Rusch was interested in 4-dimethylaminoazobenzene (DAB) as well as a number of other chemical carcinogens. Miller's research focused on the protective effect of diet and the vitamins riboflavin and pyridoxine on rat liver cancer caused by DAB.

As indicated earlier in the preceding chapter, Elizabeth was working in both biochemistry and home economics. She was dissatisfied with this combined course of study. More than eight million American men were in the armed forces toward the end of World War II, and thus there was a paucity of students in various laboratories. Because of this wartime dearth of laboratory workers and certainly a personal wish, James suggested to Elizabeth that she talk with Carl Baumann about getting into the Baumann laboratory. James A. Miller interceded on Elizabeth's behalf, and with his help Elizabeth became a valuable part of Baumann's laboratory team. While Elizabeth was housed in biochemistry with Baumann's group, she was expected to retain her association with home economics, which indeed she did. Bread, however, was no longer her primary focus. Since *both* James and Elizabeth Miller obtained their Ph.D. degrees through Baumann, and because academic heritage is so important to most scientists, it is appropriate to consider Baumann here.

Carl August Baumann: Chemist, Hero and Major Professor

Baumann, a native of Milwaukee, Wisconsin, initially enrolled at the University of Wisconsin-Milwaukee Campus but soon left for Madison, where he received his B.S. in chemistry in 1929. He stayed on at the University of Wisconsin for graduate study in biochemistry with Harry Steenbock, who later would serve

FIGURE 7.2 *Carl August Baumann, major professor to both Elizabeth and James A. Miller.*
COURTESY OF THE UNIVERSITY OF WISCONSIN-MADISON ARCHIVES, IMAGE #S11213.

on both Elizabeth's and James's thesis committees. Steenbock was studying vitamin A at the time. It was Baumann's job to investigate the relationship of carotene to vitamin A. He was awarded the Ph.D. in 1933.

The newly minted Ph.D., Carl Baumann, began a two year study in Europe, which included significant time with Richard Kuhn in Germany. Kuhn was professor at the University of Heidelberg and Director of the Kaiser Wilhelm Institute for Medical Research in Heidelberg (known, since 1950, as the Max Planck Institute). In 1938, not long after Baumann's visit, Kuhn was awarded the Nobel Prize in Chemistry for his work with carotenoids and vitamins. Kuhn was forbidden by Adolph Hitler, Chancellor of Germany at that time, to accept the Nobel Prize. However, after the demise of Hitler and the end of World War II, Kuhn received the Nobel diploma and medal but not the monetary award. The delay in the award is a reminder of the very difficult days just prior to the outbreak of World War II in Europe in 1939.[20] Baumann's Heidelberg experience with Kuhn (which was to prove useful near the end of World War II) was followed by visits to Cambridge University in England and the Carlsberg Laboratory in Copenhagen. Upon Baumann's return to Madison, he received an appointment in biochemistry at the University of Wisconsin-Madison. He was an associate professor in 1941 by the time World War II began for Americans.

Carl August Baumann's contribution to the American war effort would make a truly exciting movie if someone in Hollywood desired to portray a real scientist doing something truly useful and heroic for his country during wartime.[21] Baumann was conducting a class at the University of Wisconsin-Madison in 1944 when someone at the War Department (now known as the Department of the Army, a component of the Department of Defense) called from Washington, D.C.

---

20 The 1935 Nobel Peace Prize was awarded to Carl von Ossietzky while he was incarcerated in a German concentration camp. He was held because of his "radical pacifist" views. The notion that a prisoner of the Nazi state could be awarded a Nobel Peace Prize so infuriated Adolf Hitler that he issued a decree on 31 January 1937 that forbad German citizens from accepting *any* Nobel Prize. The decree affected Adolf Butenandt (awarded the Nobel Prize in chemistry in 1939 for his studies on sex hormones), Gerhard Domagk (awarded the Nobel Prize in Physiology or Medicine in 1939 for his studies on the sulfa drugs which were extraordinarily useful in bacterial infections prior to penicillin) as well as Richard Kuhn.

21 It is perhaps a paradox that John "Duke" Wayne, who never served in the military except on screen, was awarded the Congressional Gold Medal in 1979 and the Presidential Medal of Freedom in 1980. I doubt if anyone in Hollywood ever heard of, or cared about, the *authentic heroic activity* of Carl August Baumann.

Baumann was asked if he could read, speak, and understand German. Of course he could. He was not only born and grew up in a German-speaking neighborhood in Milwaukee, but he had had the enriching experience with Kuhn at Heidelberg's Kaiser Wilhelm Institute. The Army sought Carl because of his German cultural and language skill and because he was a superb chemist. These were sorely needed qualifications for an American top-secret operation. The military needed Baumann to be dropped *behind* enemy lines to make contact with German scientists. He was charged with learning about their research programs and ascertaining the location and amounts of uranium that could be used in the production of what are now called "weapons of mass destruction". The War Department urgently needed to know how advanced German scientists were in the field of nuclear fission and the development of a nuclear bomb. Further, additional uranium was needed in the American effort to develop a functional atomic bomb. This was extremely hazardous duty which involved many situations dangerous to the life of young Baumann. After serving his country and the allied cause with great valor, Baumann was decorated by the United States government with the Medal of Freedom. He was also awarded the Order of the British Empire (OBE) by our then closest wartime ally.[22] He returned to the University of Wisconsin-Madison when the war was over.

Back in Madison, Baumann resumed his interest in Vitamin A (and other vitamins) and began a study of the influence of diet on the metabolism of cancer cells and chemical carcinogenesis. Baumann successfully guided more than 50 graduate students to the Ph.D. degree including both James Alexander Miller and Elizabeth Cavert. Their relationship at the University of Wisconsin-Madison follows.

James Meets Elizabeth

Jim was a biochemistry teaching assistant at the time of the founding of the McArdle Laboratory in 1940.[23] Jim taught the main biochemistry laboratory

22 More information about Baumann and Operation Alsos can be found in: Suttie, J.W., P.B. Swan. 2004. Carl August Baumann (1906–1999). *Journal of Nutrition* 134:1010–1012; Atomic Rivals and the Alsos Mission, The Manhattan Project, Office of History & Heritage Resources, US Department of Energy at https://www.osti.gov/opennet/manhattan-project-history/Events/1942-1945/rivals.htm, accessed 11 August 2015.

23 I, the author, was a laboratory instructor (technically a "teaching assistant") during my graduate career and so too were my own graduate students during their graduate years.

in the first semester and he taught the laboratory on blood and urine analysis as well as vitamins in the second semester. He held his teaching assistantship for three years for which his stipend was increased from $700 at the beginning to $1,000 his last year. Elizabeth Cavert held this same teaching position for another two years after Jim completed his stint at the job. In the autumn of 1941, James had the immense good fortune of meeting Elizabeth in that laboratory course. As a student Elizabeth had a way of revealing her total attention to issues in the laboratory and completing her work with precision and dispatch. Miller wrote, "To say that I was impressed is an understatement. I paid a lot of attention to her and got to know her well."

In a short time James "fell deeply in love with Elizabeth". Jim certainly recognized how bright Elizabeth was. After all, she was a WARF scholar just as he was. She was a diligent worker, he noted, as her laboratory instructor. And, being human and male, he could not have missed how pretty she was. It was probably a combination of all of these traits, and much, much more, that attracted Jim to Elizabeth—permanently. It should be mentioned that Elizabeth was engaged to her Minnesota boyfriend when she arrived at Wisconsin and during her early time in this laboratory class. While James was shocked to learn of Elizabeth's engagement, he did not let it deter his rather significant interest in her. As he stated, he "paid a lot of attention to her". James had an immense advantage over his competition, who now lived in Iowa: James was Elizabeth's laboratory instructor and both lived in Madison. Persistence of the on site laboratory instructor and, of course, his own not insignificant qualities, led Elizabeth to agree to dates. The relationship evolved rapidly, and by summer 1942, Elizabeth broke her engagement and agreed to marry James—he later said that this was the "most important event in his entire life."

The timing of this relationship was synchronous with the onset and early months of World War II. Young people, especially men, had no idea where they would be a few months hence. Elizabeth's sister and her husband, shortly after their wedding, were separated for 42 long months as Jack Mitchell served in the Navy in the Pacific area of operations. Elizabeth's brother Mead Cavert was separated from his sweetheart from 1943 to 1946 as he served in the U.S. Air Corps in the China Burma India Theater of Operations. The young couple *may*

The position of teaching assistant provides the department with cheap labor on the one hand, but it also gives invaluable experience to the graduate student on the other hand. One cannot explain a complex scientific concept to beginning students if one has not already mastered the material—thus, the position of teaching assistant benefits both the department and the graduate student—and, even the lowly undergraduate.

have thought "we know nothing at this point about what will happen later; we can get married now—let's do it." Matrimony followed without delay.

Love, courtship, and marriage did not slow Jim in his pursuit of the Ph.D. which he received in 1943 for his research on the fluorescent properties of polycyclic aromatic hydrocarbons (see below).

Of course James learned that Elizabeth had no enthusiasm in her joint biochemistry/home economics endeavors. Recall that unlike Minnesota that welcomed Elizabeth Cavert in the biochemistry major, Wisconsin thought it necessary (in those days) to require female biochemistry students to co-major with home economics. Because of James' immense appreciation of Elizabeth's intellectual capabilities and her technical skills, Jim was able to convince his major professor Baumann to engage Elizabeth as a member of the laboratory. She was delighted to join. James too must have been delighted for now he could see more of the woman whom he so deeply loved.

The Wedding

Elizabeth and James were married in her parent's home at 1443 Grantham Street (Figure 5.3) in Saint Paul, Minnesota on August 30, 1942. It was not a huge affair; in fact, only four people attended the wedding in addition to the Reverend Arthur H. Gilmore, the pastor of the Saint Anthony Park Congregational Church, who officiated at the wedding. The others were Elizabeth's sister and her sister's husband, Jane and Jack Mitchell, and Elizabeth's parents. Where did the newly married couple go for a honeymoon? When I asked that question of Mead and June Cavert, June replied: "To the laboratory of course!" That statement is not exactly true according to both of Elizabeth's daughters. Yes, they headed to the laboratory but not before a nice honeymoon at Interstate State Parks on the Saint Croix River near Taylors Falls, Minnesota.[24]

They Became Like the Two Strands of DNA

While small, the wedding heralded the beginning of a happy family life and a lifetime of scientific investigation for Elizabeth Cavert and James Miller.

---

24 Interstate State Parks are aptly named because the parks are located in both Minnesota and Wisconsin; they were the first state parks located opposite each other—with the same name—in two contiguous states.

Many years later, James wrote about Elizabeth: "I never cease to marvel at my good fortune in having her as a marital and research partner since our graduate student days".[25] Their "symbiotic relationship" was described as "like the two strands of DNA, each one the complement of the other."[26] The two strands of DNA metaphor is reinforced by the fact that Elizabeth and James were inducted into the National Academy of Sciences *simultaneously* in 1978. Husband and wife collaborators, both inducted into the National Academy, were and are rare indeed. During the years that I spent in the preparation of this biography, I could find no one who disputed the "symbiotic relationship" described by Potter.

The story of Elizabeth's family in the preceding chapter illustrates the fact that while her immediate family was not affluent neither was it deprived. All of the family were well educated, and all had significant careers. This was not the case with James Alexander Miller. Miller was the first of his family to graduate from a university and the first to obtain a Ph.D. His family fell on hard times during the depression, and this delayed the undergraduate and Ph.D. degrees that James sought and worked for. While their backgrounds differed, both Elizabeth and James excelled as scientists and were devoted to each other and their daughters throughout their married lives.

James was a non-smoker and had a lifetime concern for health and safety issues. He was greatly concerned about Elizabeth's smoking as described in the previous chapter.[27] Jim was successful in convincing (coaxing, imploring,

---

25 Miller, J.A. 1970. Carcinogenesis by chemicals: An overview—G.H.A. Clowes Memorial Lecture. *Cancer Research* 30:559–576.

26 Potter V.R. 1986. Years with Elizabeth and James Miller: In appreciation. *Advances in Experimental Medicine & Biology*. 197:1–9.

27 Jim's concern about smoking was several years ahead of the <u>first</u> *Surgeon General's Report on Smoking and Health*, 1964, Washington, D.C., U.S. Department of Health, Education, and Welfare, Public Health Service. Was he prescient in his fear of smoking? The answer to that question is a very emphatic "no"—*much* had been written about the hazards of smoking prior to the Surgeon General's Report. Perhaps the best known of the *very early* writings was the 1604 treatise of James VI of Scotland (also known as James I of England; this is the "King James" of an early English Bible translation). King James described smoking tobacco as "daungerous to the Lungs". More than 400 years later, Alton Ochsner of Tulane University, New Orleans, Louisiana, published seminal reports of lung cancer among smoking veterans of World War I, see for example: Ochsner, A., M. DeBakey. 1939. Primary pulmonary malignancy: Treatment of total pneumonectomy. *Surgery, Gynecology, Obstetrics* 68:435–451. See also: Müller, F.H. 1939. Tabakmissbrauch und

cajoling?) Elizabeth to quit smoking, and she did quit "cold turkey". She never smoked again. Because of this, James made a lasting and good impression on Elizabeth's father. Her father was distressed when Elizabeth starting smoking while in college and was unsuccessful in his efforts of persuading Elizabeth to quit—James, as seen, was far more successful.

James' Ph.D. Research

James A. Miller's Ph.D. thesis, dated April 28, 1943,[28] was approved the next day by his committee, consisting of Professors Carl A. Baumann, Harry Steenbock, and F. M. Strong. His dissertation, to be described below, was titled: "*The Fluorescence Properties of the Carcinogenic Hydrocarbons*". His doctoral dissertation was an account of how 4-dimethyaminoazobenzene causes liver cancer in rats.

Prior to Wisconsin, James had used the same biochemical textbook by Ross Aiken Gortner that Elizabeth had used. There were a few pages in the book devoted to the newly discovered chemical carcinogens that piqued Jim's interest. Carl Baumann worked on chemical carcinogenesis. Because space was available in Carl's lab and because of James' interest in carcinogenesis, he went to work on the production of liver tumors in rats fed certain carcinogenic dyes such as the 4-dimethylaminoazobenze previously mentioned.

Concerning James A. Miller's dissertation, a key word is the adjective "carcinogenic", which means cancer-forming. A hydrocarbon is an organic substance containing only hydrogen and carbon. Of course, fluorescence refers to the emission of light by a substance stimulated by radiation such as ultraviolet light. Jim was studying fluorescence in the ultraviolet of cancer-causing organic chemicals.

Lungenkarzinom. *Zeitschrift für Krebsforsschung* 49:57–85. A cursory examination of the history of tobacco use will reveal many more pre-Surgeon General studies. James Alexander Miller devoted his life to cancer research—even as a graduate student, it is entirely reasonable to expect that he knew of these publications and thereby understood the hazards of tobacco.

28 James Alexander Miller was at Wisconsin during the war years. How did he escape the wartime draft? According to Jim (Oral History, U of Wisconsin), he sought the service of Nathan Feinsinger (1902–1983), Professor of Law at Wisconsin, who negotiated with the Pittsburgh draft board to obtain deferment for Jim because of his scientific work.

In his thesis introduction James stated that much desirable information was not yet available concerning chemical carcinogens. Considered "desirable information" are what property of the chemical causes changes in cells resulting in cancer and what fate awaits the carcinogen in the body. James Alexander Miller and his spouse focused their entire lifetime research on these questions, and for James it started early with his thesis research.

It was already known that certain hydrocarbons, known as polycyclic aromatic hydrocarbons (PAHS), caused cancer. What was needed was a method for quantitative detection of the carcinogenic hydrocarbons and their metabolic breakdown products in the tissues of treated animals. Jim studied a number of factors that might affect the fluorescence of the chemical under consideration and then did tissue analysis. His thesis concluded with a review of several concepts explaining how chemical carcinogens wrought their havoc in living cells. As stated in the previous paragraph, as early as 1943, James was considering scientific issues that would concern Elizabeth and him for the remainder of their professional careers.

In 1944, James Alexander Miller, Ph.D. was recruited to the newly founded McArdle Laboratory for Cancer Research by its first director, Dr. Harold Rusch. James remained at the McArdle throughout his entire scientific career. So too did Elizabeth.

Elizabeth's Ph.D. Research

James thesis was followed two years later, on April 23, 1945, by Elizabeth's Ph.D. thesis "*The Effect of Pyridoxine and its Derivatives on Growth and Xanthurenic Acid Excretion*". She was then only 24 years old.[29] With reference to Elizabeth's thesis, some people will recognize that pyridoxine is another name for the water-soluble vitamin B6. Pyridoxine is found in most foods, such as meats, vegetables and whole grains. Pyridoxine plays a role in the metabolism of amino acids in general and the essential amino acid tryptophan in particular. A metabolic end product of tryptophan is xanthurenic acid found in the urine. Elizabeth was studying the toxic effects of high-protein diets in mice deficient in pyridoxine, and the role of this deficiency on growth and the excretion of

29 Elizabeth C. Miller's Ph.D. thesis is dated April 23, 1945. She became 25 on May 2, 1945. This is a relatively young age to finish the Ph.D. It may be compared with a recent study of average age at the doctorate for biological sciences of 30.8 years and for chemistry of 29.6 years; see Hoffer, T.B., V. Welch Jr. 2006. *Time to Degree of U.S. Research Doctoral Recipients*. National Science Foundation. NSF 06-312.

xanthurenic acid in mice. Her thesis research concerning this function of pyridoxine in tryptophane metabolism was cited in biochemistry textbooks and at least 113 journal articles—long before computer generated bibliographies.

What one learns, from reading either the thesis or the published scientific reports derived from thesis data, is that Elizabeth read and knew the relevant literature. The bibliography of cited scientific publications in her thesis is five and a half pages long. More importantly, Elizabeth innovated quantitative chemical methods for the determination of xanthurenic acid in the urine of mice and rats. She maintained her animals in such a manner that she controlled their diet and captured their urine with its xanthurenic acid. She was then able to correlate dietary levels of casein (a milk protein), tryptophan and pyridoxine with growth of the animals and excretion of the tryptophan end product. Finally, Elizabeth's research reported the effects of diet to modulate the competence of azo dyes (e.g., p-dimethylaminoazobenzene, DAB) to induce rat liver cancer. A note should be made that both James and Elizabeth worked on the liver-cancer-inducing chemical DAB; their life in science had a commonality of the same university, same department, same major professor and even same cancer-causing chemical in their doctoral dissertations. These commonalities resulted in the abundantly productive careers that gave rise to this book.

Years later when asked how she got into cancer research, Elizabeth replied that it was a consequence of starting her career in Carl Baumann's laboratory. The nutritional effect of a vitamin on the competence of a chemical carcinogen to cause cancer was her start.[30] Elizabeth's thesis was in keeping with the preeminence of the University of Wisconsin-Madison in vitamin research; researchers included Steenbock on vitamin D, Elvehjem's anti-pellagra factor niacin, and Baumann on vitamin A, pyridoxine, and folic acid. Later, Elizabeth and James focused on how chemical carcinogens became metabolically activated and ultimately reacted with and bound to large molecules of normal cells resulting in cancer.

In more general terms, and especially in context with her subsequent career, I think that her thesis research supports the view that she could and did master the scientific literature, she *innovated* quantitative chemical procedures, she studied metabolism as it relates to diet and other factors, and finally, she correlated this kind of information as it affected chemical carcinogenesis. All of this was used productively throughout her career. While her academic career

30 There is an extensive literature of the effect of vitamins and other dietary factors on carcinogenesis by aminoazo dyes; see Miller, J.A., E.C. Miller. 1953. The carcinogenic aminoazo dyes. *Advances in Cancer Research* 1:339–396.

began as an undergraduate in *agricultural* biochemistry, she is remembered as a giant in the field of *chemical carcinogenesis*.

Elizabeth lost no time in converting her thesis data into published papers. All or certainly most of Elizabeth Cavert Miller's scientific publications can be accessed by referring to the appendix in this book. However, it is perhaps more convenient to list her thesis publications as a footnote on this page.[31]

Elizabeth used small animals in her research. These were mostly rats and mice. There was then, and there are now, no computer program or *in vitro* cell strain that could provide answers for her research. While some well-intentioned, present-day students promote animal rights and lament the use of animals for research, Elizabeth used them without apology.

I would like to make the point that seldom does only *one* professor guide a graduate student toward the doctorate. Other professors (and other graduate students) serve as backup resources for technical information, procedures and provide friendly critical evaluation of experiments and the writing of the resulting doctoral dissertation. In addition to this informal backup system, there is a formal committee that aids and assists (and tests) the student. Elizabeth was enriched by the committee who "signed off" on her thesis. Leading that list of three was Harry Steenbock, already mentioned with reference to the Wisconsin Alumni Research Foundation (WARF) and the ultraviolet irradiation of milk, followed by Alfred Lawrence Wilds (1916–2002) and Conrad Arnold Elvehjem (1901–1962). Wilds was an organic chemist who studied steroid hormone chemistry.[32] Elvehjem was an agricultural chemist interested in vitamin and mineral nutrition. He was perhaps best known for his isolation of niacin in the late 1930s. When Elizabeth received her Ph.D., Elvehjem was chair of biochemistry, after which he served as Dean of the Graduate School and then President of the University of Wisconsin-Madison. This committee of three professors plus Carl A. Baumann (who was on academic leave when

---

31 Miller, E.C., C.A. Baumann. 1945. Relative effects of casein and tryptophan on the health and xanthurenic acid excretion of pyridoxine-deficient mice. *Journal of Biological Chemistry* 157:551–562. Miller, E.C., C.A. Baumann. 1945. Further studies on factors that affect xanthurenic acid excretion. *Journal of Biological Chemistry* 159:173–183. Miller, E.C., C.A. Baumann, H.P. Rusch. 1945. Certain effects of dietary pyridoxine and casein on the carcinogenicity of *p*-dimethylaminoazobenzene. *Cancer Research* 5:713–716.

32 Steroids are complex organic chemicals that are included with the lipids (fats and fat-like substances) because of their solubility in fat solvents. Cholesterol, sex hormones, bile acids, anabolic steroids, progesterone and adrenocortical hormones are examples of steroids, as is vitamin D.

Elizabeth finished her degree) was obviously a stellar group to guide and assist the young Elizabeth.[33]

James and Elizabeth at the McArdle

Established in 1940, the McArdle Laboratory for Cancer Research,[34] (see Figure 3.1) was an early research institute devoted to cancer.[35] The McArdle was the first such institution to be founded within a university. Dr. Harold P. Rusch (1908–1988) was its first director. The McArdle grew from a modest beginning to an institution that now has a staff of approximately 200. James Miller was the third member of the laboratory staff, Rusch and Potter[36] having been numbers one and two. Now the faculty directs 22 research groups. Since its founding, more than 1,800 pre-doctoral and post-doctoral students have passed through its laboratory doors.

During their tenure at the McArdle the Millers had close contact with many students. Elizabeth served not only her laboratory and the students therein, but she was also a long- time associate director. She was described as always available for consultation and she would put aside papers on which she was working to provide help and advice to a student in need. Indeed, Fred

33 Elizabeth's doctoral thesis lists "Home Economics" as a minor, but no person from that department was on her doctoral committee.

34 The McArdle Laboratory for Cancer Research at the University of Wisconsin was named for Michael W. McArdle (1874–1935), a president of the Sunbeam Corporation and a Chicago attorney. McArdle bequeathed money, which with matching funds from the Public Works Administration (PWA), was used in 1940 to construct the McArdle Laboratory for Cancer Research; see: Shimkin, M.B. 1979. Eponymic cancer institutes in the United States. *Cancer Research* 39:2211–2215.

35 Other early cancer research laboratories include the Institute for Cancer Research located in Fox Chase, Philadelphia, Pennsylvania. The Institute was founded in 1927. It merged with the American Oncologic Hospital in 1974 to form the Fox Chase Cancer Center. The Sloan-Kettering Institute, a part of the Memorial Sloan-Kettering Cancer Center, was founded in 1945. The Sloan-Kettering Institute is located at 1275 York Avenue, New York City.

36 Van Rensselaer Potter (1911–2001), a native of Day County, South Dakota, graduated from South Dakota State College (now South Dakota State University) and received his Ph.D. in 1938 from the University of Wisconsin working with Conrad Elvehjem. Potter published papers in the area of enzyme biology and nucleic acids. He is remembered for his dictum, "oncogeny is blocked ontogeny", referring to the failure of tumor cells to attain maturity, and to his coining the term and writings in "bioethics". He was a Professor of Oncology at the University of Wisconsin Madison.

Kadlubar stated, about both Elizabeth and Jim, that "their willingness to stop what they were doing and talk about science at any time, weekdays or weekends" was an unforgettable quality: a student "never had to make an appointment and their door was always open."[37] Jim showed his interest as a mentor in perhaps a more practical way. Allan H. Conney[38] wrote that his work place in the Miller laboratory was often a mess. Jim suggested gently several times that Allan ought to clean up his work place, but Jim's request came to naught. Allan returned to the laboratory one Saturday morning and "was greatly distressed to find Jim cleaning my glassware and lab bench. When I tried to take over the glassware cleaning job, Jim simply said that he could handle it himself. This approach to teaching good housekeeping practices made a lasting impression". Another graduate student in the laboratory, Norman Drinkwater, described Jim as "extremely patient". And, he was "a truly exceptional mentor, a wise and faithful counselor and teacher who brought out the very best in each of us". Drinkwater acknowledged that he and two others came near to blowing up the lab at one particular time. Jim's comment at this near disaster was only that "it really is a good idea to check Fieser before setting up that sort of thing".[39,40] Yet another student Miriam Poirier stated "we graduate students learned experimental creativity, hypothesis validation, and a scientific caution ... (his) exquisite sense of optimal experimental design is among the most valuable gifts that Jim gave to his students, and for that I have been profoundly grateful every day of my research career." Poirier, who received her Ph.D. elsewhere, commented further that "Betty and I frequently met at meetings, and I always

---

37 Kadlubar, Fred F., Ph.D., University of Texas Austin; Postdoctoral research, McArdle Laboratory for Cancer Research, University of Wisconsin-Madison. Kadlubar, now deceased, was Chair, Department of Molecular Epidemiology, School of Public Health, University of Arkansas School of Medical Sciences, Little Rock, Arkansas.

38 Conney, A.H., M.C. Poirier, Y.-J. Surh, F.F. Kadlubar. 2009. Elizabeth Cavert Miller (1920–1987) and James A Miller (1915–2000): A Biographical Memoir. Washington, D.C.: *National Academy of Sciences*.

39 Drinkwater, Norman. Ph.D., University of Wisconsin-Madison, 1980. Marian A. Messerschmidt Professor in Cancer Research, University of Wisconsin-Madison. see pages 15–16: Conney, A.H., M.C. Poirier, Y.-J. Surh, F.F. Kadlubar. *Elizabeth Cavert Miller 1920–1987, James A. Miller 1915–2000, A Biographical Memoir*. 2009. Washington, D.C.: National Academy of Sciences.

40 The "Fieser" mentioned was likely Fieser, L.F., M. Fieser. 1967. *Reagents for Organic Synthesis*. New York, John Wiley and Sons, Inc.

delighted in her wisdom and friendship. Jim's support was something I enjoyed and treasured for the remainder of his life."[41]

James and Elizabeth Miller were recruited to the McArdle Laboratory for Cancer Research by Harold P. Rusch. Jim had an earlier start by working for a time on chemical carcinogenesis with the founder Rusch; he served as a Finney-Howell postdoctoral fellow after which he became an instructor in 1944. Elizabeth, like Jim, became a Finney-Howell postdoctoral fellow but later in 1945; she served two years before becoming an instructor in 1947. James and Elizabeth became Professors of Oncology in 1969, then Van Rensselaer Potter Professors of Oncology in 1982, and finally WARF Senior Distinguished Professors of Oncology in 1984. Elizabeth and James shared a 45-year joint career at the McArdle. Elizabeth became Professor Emeritus June 1, 1987. James became Professor Emeritus two years earlier, June 7, 1985. Elizabeth was Associate Director of the McArdle from 1973 until the time of her death in 1987. She served during the period when Henry C. Pitot[42] was Director. The director had the choice for his associate director. Pitot chose Elizabeth Cavert Miller and concerning this choice, he stated, "One makes few really outstanding correct decisions in one's time on this earth. The decision to ask Betty Miller to assist me in the administrative duties of the McArdle Laboratory was one of the best decisions I have or ever will make."[43] According to Pitot, Elizabeth viewed her administrative responsibilities as service to her colleagues—she was concerned that scientific studies of the McArdle be of the absolute highest quality. She strove for the very best from herself and all of her colleagues at the McArdle. Sometimes administrators are viewed by the staff as "paper pushers." No "paper pusher" was Elizabeth Cavert Miller.

---

41 Poirier, Miriam Christine Mohrhoff, Ph.D., 1977, Catholic University, Washington, D.C. Current position is Head, Carcinogen-DNA Interactions Section, National Cancer Institute, National Institutes of Health, Bethesda, Maryland.

42 Henry Clement Pitot, Professor Emeritus of Oncology and of Pathology and Laboratory Medicine, University of Wisconsin. Pitot was born in New York City, May 12, 1930. He received his M.D. and Ph.D. degrees from Tulane University, New Orleans, Louisiana. He was Director of the McArdle Laboratory for Cancer Research from 1979 through 1991. Pitot was first a member (1976–1979) and then Chairman (1979–1982) of the President's National Cancer Advisory Board.

43 Conney, A.H., M.C. Poirier, Y.-J. Surh, F.F. Kadlubar. 2009. Elizabeth Cavert Miller (1920–1987) and James A. Miller (1915–2000): A Biographical Memoir. Washington, D.C.: *National Academy of Sciences*.

CHAPTER 8

Elizabeth and James, Beyond the Lab

Like the Two Strands of DNA—But with Differences

Elizabeth and Jim Miller's overall approach to life was similar. They shared a passion for work, family, science and each other. They were together almost all of the time and disagreements were uncommon. They meshed together so well that they were described as "like the two strands of DNA—each one the complement of the other."[1] While this partnership is certainly a fact, as individuals they could be quite different from each other. It is a tribute to the force of both of their personalities and their love for each other and their children that they could blend their differences into a smoothly working team. As the daughters related to this author, "They cared deeply about creating a solid and caring family life."[2]

"They Seemed Normal to Us"

After a decade of marriage devoted to launching their careers in cancer research the Millers had two daughters, Linda and Helen. As children Linda and Helen did not realize that their parents were unusually talented and productive scientists. As they professed, "they seemed normal to us—they were just Mom and Dad." But being "just" Mom and Dad did not mean that the girls failed to note that both parents worked. Most mothers were stay-at-home moms at the time, and they correctly surmised that both parents working was not the social norm.[3]

1 Potter, V.R. 1986. Years with Elizabeth and James Miller: In Appreciation. *Advances in Experimental Medicine and Biology* 197:1–9.

2 *Jim and Betty Miller—Our Parents*. 2010. Forbess, L.M., Alexander, H.M. Typescript personal communication. This document, plus personal interviews, are the principal sources of information about the Miller family.

3 During the first half of the 1950s, less than 25% of married mothers with children worked. By the mid-1990s, about 70% of such women were employed. See: Cohany, S.R., E. Soc. 2007. Trends in labor force participation of married mothers of infants. *Monthly Labor Review* 130:9–15.

 | DOI 10.1163/9789004286801_009

Jim regarded Elizabeth as a "conscientious mother" as did the Millers' colleagues at the McArdle Laboratory. So too did the daughters. Helen and Linda told Roswell K. Boutwell that when they were young, they thought of Elizabeth as a typical mother. Elizabeth did all the things that Linda and Helen expected of a mother—and much more. After the girls were in bed and were thought to be asleep, Elizabeth resumed her scientific work at the typewriter.[4]

A personal note is in order here. I had the responsibility of running a university laboratory for almost forty years. I became a father of three children during that time. I am forever grateful that I had a wife who assumed the lion's share of childhood care. Unquestionably, my wife was also a "conscientious mother." In this context, I cannot imagine how Elizabeth handled pregnancy, child care, attention to her husband, meal preparation and certainly not least, her professional career. I cannot imagine how she juggled all of these tasks and remained scientifically productive, sane, and a "conscientious mother" simultaneously. It is called multitasking now and obviously she was a magnificent master multitasker. My admiration for her personal life is unbounded. Nobel Laureate Howard Temin (1934–1994) voiced a similar notion when commenting on Elizabeth's cancer research and family. "In her life, Betty accomplished an incredible amount in cancer research and at McArdle in her quiet, efficient, unflappable way. She did all this while raising two delightful daughters and maintaining a close family life and home—a difficult juggling act for anyone." Temin also wrote that "Betty had an intense institutional loyalty to McArdle, the University of Wisconsin-Madison, Jim, her family, cancer research, and science in general."[5]

---

4 Boutwell, RK is a distinguished professor of oncology emeritus at the McArdle Laboratory. He was born in Madison, Wisconsin 24 November 1917 and obtained his Ph.D. in biochemistry from the University of Wisconsin. See his comments in: Conney, A.H., M.C. Poirier, Y-J. Surh, F.F. Kadlubar. 2009. Elizabeth Cavert Miller 1920–1987, James A. Miller 1915–2000: A Biographical Memoir. Washington, D.C.: *National Academy of Sciences*, page 12.

5 Temin, Howard Martin (1934–1994), born in Philadelphia, Pennsylvania and died in Madison, Wisconsin. Temin shared the Nobel Prize with David Baltimore for the discovery and characterization of reverse transcriptase (RNA-dependent DNA polymerase). Reverse transcriptase is an important enzyme in (among other diseases and conditions) AIDS, hepatitis B, and maintenance of chromosome telomeres. His comments concerning Elizabeth Cavert Miller are found in Conney, A.H., M.C. Poirier, Y.-J. Surh, F.F. Kadlubar. 2009. Elizabeth Cavert Miller 1920–1987, James A. Miller 1915–2000. Washington, DC: National Academy of Sciences, page 12.

Elizabeth and James' Daughters

Linda Ann Miller was born in 1952 and is now Linda Miller Forbess, who lives in Schofield, Wisconsin. Linda graduated from the University of Wisconsin-Madison and obtained her master's degree from the University of Wisconsin-Menomonie in textile design. She has done doctoral work in the History of Design. Linda is an expert on fabrics and the design of clothing. She has businesses known as "Seams Old" and "Seams Perfect". She is the Curator of Textiles at the Marathon County (Wisconsin) Historical Society.

The second daughter, Helen Louise Miller, was born in 1955 and also obtained her undergraduate degree from the University of Wisconsin. Her mother chose the University of Minnesota because it was "close and good". Helen chose Wisconsin for essentially the same reason—the Madison campus was close and good (and, just as the author upgraded the term "good" for the University of Minnesota, I would similarly change "good" to "outstanding" in describing the University of Wisconsin Madison). Upon completion of her undergraduate

FIGURE 8.1 *Jim and Elizabeth Miller at the festive occasion of daughter Helen's wedding, Left to right: Linda Miller Forbess, James A. Miller, Helen Miller Alexander, David Alexander, Elizabeth C. Miller, Russ Forbess.*
PHOTOGRAPH COURTESY OF THE MILLER FAMILY.

studies, Helen knew that she wanted to attend graduate school and thus applied for admission at Duke University as a master's degree candidate. She was told at her admission interview that Duke admitted few master candidates and that her undergraduate grades were entirely suitable for admission to their Ph.D. program. Helen altered her application right then and there to indicate that she would study for the Ph.D. Helen received this degree in botany in 1982 from Duke University. Helen married David Alexander[6] and is now a professor in the Department of Ecology and Evolutionary Biology at the University of Kansas, Lawrence, Kansas. In addition to teaching, she studies the population biology of plants. Helen Alexander has won numerous awards for her teaching at the University of Kansas and in 2007 she was inducted into the University of Kansas Women's Hall of Fame (see http://www2.kumc.edu/etwrc/).

Elizabeth Cavert Miller as "Mom"

Elizabeth was well organized both at home with family and at the laboratory. She had to be. She was constantly in motion, focusing on the job at hand and planning ahead for the next project. While she was an extraordinarily creative and productive scientist, this in no way took precedence over her family and home responsibilities.

Elizabeth was a woman of her social time and upbringing. In mid-20th century, women were expected to take the primary role of parenting and running the home. This was the role for a woman even if she worked outside of the home. Elizabeth, with all of her laboratory duties, did not shy away from her domestic responsibilities. In fact, she chose to be employed part-time while the girls were young, so she could leave the university as needed to help her children. The reduced pay for less than full-time work was, in fact, a partial fiction; since she often worked after the girls were in bed, she most likely put in 40+ hour weeks. Curiously, but perhaps not surprisingly, Elizabeth accomplished as much research and administration during the half-time appointment years working at home as she did with her previous full-time appointment.

Mrs. Margaret Davis had something to do with Elizabeth's productivity after the arrival of the children. Mrs. Davis was hired as a housekeeper/babysitter and worked full-time while the daughters were small, then part-time when

6 Dr. David E. Alexander is a member of the Department of Ecology and Evolutionary Biology, University of Kansas, Lawrence, Kansas. David Alexander is interested in, among other topics, the biomechanics of flying animals.

the girls were growing up until she retired. Mrs. Davis was married and had a daughter of her own.

Mrs. Davis was at the Miller home when Linda and Helen came home from school. She relieved Elizabeth of cleaning and laundry chores and quite frequently started the evening meal prior to the return of Elizabeth and Jim. Elizabeth would finish the cooking after Mrs. Davis left for the day. Elizabeth made certain that the girls understood that Margaret Davis was not a servant and that she had her own work to do in the Miller home. Linda and Helen were not to hamper that work. Mrs. Davis was instructed by Elizabeth not to wait on the children. The girls had their own chores. Elizabeth expected them to set the table for dinner, pick up the dishes and place them in the dishwasher after dinner. On weekends, there were outdoor chores assigned to the girls. Elizabeth was careful to teach the children to address Mrs. Davis as just that; no child in the Miller family was going to call her "Marge". Children addressed adults by title and last name as a sign of respect, and Mrs. Davis had clearly earned respect.[7]

While Davis was a self-described nanny/housekeeper, she was also a friend and in some ways a second mother. She played with Linda and Helen when they were children. Linda and Helen made occasional trips to see Mrs. Davis even after they were grown and had left home. Mrs. Davis told Helen's children that Linda and Helen were *perfect* and well behaved as children. Linda and Helen both deny being "perfect" but the description of two little girls by Mrs. Davis shows the affection that this devoted woman had for the Miller family. Although Mrs. Davis was obviously of great help, Elizabeth could not and did not shun her familial responsibilities.

The Millers at Home

Elizabeth and Jim left home together for the laboratory in the morning and came home together in the evening. Back in those days of the 1950s and 1960s, the family ate together and that is no longer the social norm.[8] Dinner conversation ranged from what the girls had done in school on that particular day or what their friends were doing to the more arcane subject of laboratory

7 Mrs. Davis worked for the Millers in mid-20th century. Customs evolve. In early 21st century, my Minneapolis grandchildren addressed their school teachers by their first names. Had they known Mrs. Davis in the early 21st century, they would probably have addressed her as Margaret—with no disrespect intended.

8 Fieldhouse, P. 2007. Eating together: The culture of the family meal. *Transition* 37:3–6.

research. Elizabeth and Jim would laugh when the girls teased them about using seemingly inscrutable polysyllabic biochemical terms. Note however, the fortunate daughters were not excluded from technical talk and were thus exposed to scientific terminology, and more importantly, scientific thinking and commitment at an early age.

The after-dinner hours were a time for the girls' homework and a time for Jim and Elizabeth to do work-related reading or writing. Concerning the homework, Linda and Helen grew up prior to the days of laptop computers and the ease of finding information. So, the girls would seek information from Mom and Dad as necessary. However, Linda and Helen preferred to get their facts and figures from their mother more than from their father. Elizabeth was a patient mother and always made time for the daughter who needed help. When asked, it appears that she knew just how much her daughters needed and she could be counted on to give a short and concise answer; just enough to solve the problem needed or provide background information for a school paper. This is not to say that she would not spend considerable time with a daughter when necessary. Linda recalls receiving from her mother hours of assistance in grade school as she endeavored to learn how to write a summary of a subject. The girls remember that Jim was also eager to help. However his scientific background and natural curiosity led him to probe deeper into a question than the girls wanted or needed; thus, Mom was the expedient choice for homework help.

Elizabeth read to the girls at bedtime. Elizabeth must have been a woman of boundless energy or, as the girls noted, a quintessential "super mom". After bedtime, when the girls were supposed to be asleep, and when Elizabeth had earned a well-deserved break, Linda and Helen would hear the tap-tap-tap of the typewriter as she resumed her scientific work. This was, as noted above, prior to laptop personal computers. Editing of a scientific manuscript by Elizabeth *literally* consisted of cutting and pasting. When personal computers became common, Elizabeth lost no time in learning modern word processing. Helen said that her mother seemed not to desire "down time" after her she attended to her domestic responsibilities. Elizabeth was very conscious of her laboratory responsibilities and deadlines. So, it was back to work with the tap-tap-tap when the kids were in bed.

On weekends, Elizabeth worked in the yard. She planted flowers. She set out tomato plants. She trimmed the bushes. She mowed the yard. She raked the leaves. Helen (much as most kids did) thought it was great fun to look at and play near the leaf fire (in a time prior to the anti-burning ordinances of the late 20th Century). Jim assumed some of the yard care in later years, perhaps in part because Elizabeth believed he needed the exercise for health.

The Millers were a one-car but two-lawn mower family. The multiple mowers related to Elizabeth's concern for Jim's health. Concern or not, the extra mower remained idle most of the time. This is not to say that Jim was uninterested in the lawn. Actually, he hated dandelions and paid his daughters one cent apiece to dig them. The pay scale was somewhat reminiscent of the nickel an hour that Elizabeth and her brother Mead were paid to help their father grow beans and turkeys in the Depression years. A difference emerged—Elizabeth stuck to her nickel an hour job; the Miller girls, in contrast, balked at the paltry pay and only worked one time. Jim's impassioned hatred of dandelions persisted sans abatement. I asked Helen if her mother and father had a summer cottage on a lake as so many Minnesota and Wisconsin families do. Helen said that her mother believed that one home to take care of was enough and so the answer to my question was a simple "no".

Elizabeth taught her daughters how to swim and skate, played games with them and took them shopping. Daughter Linda was coached on how to swing a baseball bat. Elizabeth became co-leader of Linda's Girl Scout troop. Like many a mom of a Girl Scout, Elizabeth did her part by buying generous supplies of Girl Scout cookies— frozen for consumption at a later date. Linda had a nascent interest in sewing, knitting, embroidery, fabrics and other weaving crafts. Elizabeth had little skill and interest in such endeavors, but despite this apparent inadequacy, she willingly helped Linda's interest grow and blossom. Linda's interest was not ephemeral—she now has businesses in Schofield, Wisconsin as cited earlier.

The family also had pets. Linda had a cat when she was a little girl and quite naturally became very upset when the cat died. Linda knew her mother was not particularly interested in pet animals other than to be certain that they were fed and cared for. However, Linda's dad was fond of the animals in the home and would talk with them and pet them. Linda took her grief to her dad with a request. Dad agreed that she could have a new cat. Only then did Linda approach her Mom with the news that Dad had already approved the acquisition of another cat. When Helen was yet a girl, she urged her parents to buy a dog. In time a black, medium-sized poodle showed up at home. The poodle was evidently blessed with a long life expectancy, because Helen's mother was caring for the poodle long after Helen left home.

Of course, Mom was attentive when her daughters were ill and took the girls to the doctor's office when required. She did the driving to friend's houses and the other places that the daughters needed to go after school. She took over most of the parental tasks. Elizabeth, according to Helen, expected to be the one who did most of the "things with the children". In later years, the daughters had boyfriends (who became future husbands). Again, there were modest differences between Mom and Dad. When Russ (who became Linda's husband) or Dave (who became Helen's husband) were visiting at the home, it was Mom who provided cookies for the boys while Dad was close by, as Helen related, "reading a book".

Although Wisconsin is not part of the American "Bible Belt", it is common for many people to attend church services, and this was true for the Miller mother and her two daughters. Elizabeth's mother and father both descended from staunch Presbyterians in upstate New York. In spite of the fact that organized religion was not of great importance to her, Elizabeth wanted a church experience for her daughters and chose the First Unitarian Society of Madison. Helen believed that her mother was more "comfortable" with the liberal Unitarians than she would be with churches that were more fundamentalist. It would seem that Elizabeth wanted the liberal church experience to be a contributing factor to their "broad education".

The women of the Miller family had much to admire or enjoy with the Unitarians. To begin with, the First Unitarian Meeting House in Madison, Wisconsin was designed by the well-known architect, Frank Lloyd Wright. The extraordinarily elegant building was finished in 1951. It has been described as an "architectural treasure". The Meeting House was appropriately placed on the National Register of Historic Places (1973) and it has been designated as a National Historic Landmark (2004). Winston Churchill is credited with saying: "We shape our buildings; thereafter they shape us." Perhaps Elizabeth hoped that the Unitarian Meeting House, its congregation and its minister would, in part, "shape" her daughters in a liberal tradition. While that notion is entirely speculative, we do know that Elizabeth enjoyed helping at the Sunday School that her daughters attended. Further, and probably to Elizabeth's pleasure, Helen said that she enjoyed listening to the messages of the distinguished minister, Max D. Gaebler.[9] Dad Miller did not attend services at the Unitarian Meeting House with Elizabeth and the girls as he was proud to be a free thinker and an atheist (he sometimes used the term "agnostic").

Laboratory animals need care regardless of the day of the week. Elizabeth kindly made certain that the animal caretakers had Christmas "off". That meant that Elizabeth would drive to the laboratory, clean each cage, and make certain that there was food and water for the animals. This was done with the "help" of her small daughters and the animal care trip became an outing for the girls. The drive to the laboratory was so customary for the girls that they came to think of it as a universal experience of all children. A very little Linda

9 Max D. Gaebler (born 1921), attended the Harvard Divinity School and was the minister at the Madison church for over 35 years.

on the way to the lab with her mother once saw another car with a mother and daughter. Linda, quite naturally, thought the other car with its mother and daughter was on the way to *their* lab to care for *their* animals.[10] The lab visits on Christmas day was part of the holiday festivities for many years. Jim checked the animals occasionally, but Elizabeth did so much more frequently. Jim was thought to have had a "distaste" for that area of the Miller's research. Jim was far more interested in doing chemistry in the lab with test tubes and beakers. If Jim had a distaste for animal care, it may have had its origin many years previously when he was the animal care man for Professor Charles Glen King at the University of Pittsburgh.

Elizabeth was a manager in the home as well as in the laboratory. She kept the home going by buying the groceries, paying the bills and doing the taxes. It should be noted that the preparation of taxes required a "financial wizard" of sorts, and Elizabeth was clearly up to the task. She remained so even after the daughters left home. Linda and Helen would call on her from time to time to help with tax questions. Will Cavert, Elizabeth's father, invested wisely in the stock market. Elizabeth followed her father and similarly became interested in stocks. Elizabeth followed the market carefully and her portfolio reflected her attention to details in the market.

Elizabeth was also the girls' "employer". As Linda and Helen grew older and could assume greater responsibility, they were hired by their parents to help with the library work related to the preparation of scientific papers and to perform relatively simple tasks in the laboratory. These were "summer jobs" and the payroll for the work came out of their parents' personal income. With time, the girls came slowly to the understanding of the work of their parents and also the esteem and respect the scientific community had for them. Although the parents were modest indeed and rarely spoke of the honors, Linda and Helen were aware that the awards and plaques associated with the awards (see Appendix) were buried in a closet with other things seldom used or needed.

---

10 In a perhaps similar situation many years earlier, a Charles Darwin child, during the Darwin eight year barnacle study, was said to ask a neighbor friend about the friend's father "Then where does he do his barnacles?" as if, not only the Darwin father, but all fathers, "do barnacles". See: Darwin, F., A.C. Seward (Eds.). 1903. *More letters of Charles Darwin, A Record of His Work in a Series of Hitherto Unpublished Letters*. Volume 1. New York: D Appleton. Page 38.

The Millers Away from Home

Elizabeth and Jim had cookouts and went camping and canoeing with their daughters. Helen, by her own words, "lobbied" her parents for summer camping trips. Elizabeth was enthusiastic, but Jim was far from excited. The father had done no camping as a young boy and thus as an adult took it up rather reluctantly. The family outings were modest to begin with and close to home. Elizabeth and Jim began by taking the girls to a nearby state park during the day with a motel at night. I suppose that the motel was safer and more comfortable for sleeping, and perhaps the hot shower at the end of the day to wash away outdoor muck and mosquito repellant was appreciated.

The early trips to nearby state parks evolved into more sophisticated camping. The Millers had a canvas tent which was outfitted with sleeping bags and thin foam mats for comfort. The "comfort" of the pad was sorely inadequate for Jim's overworked back so the back problem was mitigated with additional foam. While Jim was not a veteran camper, he learned to enjoy the out-of-doors. Helen recalls his disdain for camping "slobs"—Jim carried a small bag to pick up trash that was left behind by careless hikers along the trails. Jim did not like noisy campers either but alas, he could not put them, the noisy ones, in his small bag.

Starrett Lake in northern Wisconsin was a wonderful choice as a camping site for the Millers. Noise was clearly not a factor at Starrett because no boats with motors were allowed on the lake. The campsites were surrounded by a state forest with towering pines, and "peace and quiet" was expected. Starrett Lake is located near Woodruff, Wisconsin, just a bit more than 200 miles from Madison.

Michigan's upper peninsula is the site of the Ottawa National Forest with its Sylvania Wilderness Area. The forest is due north and only a little more than 250 miles from Madison, Wisconsin. Ottawa National Forest has a million acres and surrounds the Sylvania Wilderness Area, about seven miles west of Watersmeet, Michigan. The wilderness area contains many deep and clear lakes and old growth forests in its 18,327 acres. The Millers reserved camp sites in the Sylvania Wilderness accessible only by portage, and it was here that Russell (also known as "Russ") Forbess, Linda's boyfriend and eventual husband, was helpful. The Miller family spent four or five days camping and canoeing each of the years that they were fortunate to be at the Sylvania Wilderness Area. New lakes (to them) were explored after appropriate portage. The girls picked wild blueberries that were incredibly delicious in pancakes made by their Mom over the fire.

All good wilderness areas have maps and trails. Helen was amazed that her mother did not need to follow a marked trail. She remembered at least one “compass bushwhacking” trip. Elizabeth had selected a destination and, with compass in hand, set out ignoring the trail and sans error led Helen to their selected geographic objective. However, this sort of cross-country perambulation was not necessarily typical of the camping experiences. Rather, Helen describes most of the camping trips as “pretty laid back” with enjoyment of fishing, swimming, canoeing and reading by the fire—all at a leisurely pace.

Elizabeth had an adventurous streak which was sometimes revealed when out-of-doors. On a trip to the Governor Dodge State Park located in the unglaciated area of southwest Wisconsin, a place of hills, bluffs, and deep valleys, Elizabeth climbed the beautiful rocks to better see the surrounding landscape. Helen remembers the concern of her father who would shout to Elizabeth “Get back Lizzie, Lizzie get down.” I particularly noted this recollection of Helen’s because, while I had heard Elizabeth referred to as “Betty” many times, it was the first instance that I had heard of her being called “Lizzie”. The daughters think that Jim used “Lizzie” to get their mother’s attention or when he was stressed, as well he might have been when Elizabeth got too close to the edge of the cliff.

In high school, Helen and a group of her friends planned a trip to the Florida Keys and Everglades. Elizabeth Cavert Miller volunteered to accompany the group as chaperone. Helen’s mother was an absolute essential, because the young people were only of high school age, and the presence of Elizabeth was sufficient to convince the other parents that the trip was workable and reasonable. A quick check of a map shows that the distance from Madison to the Everglades is over 1,500 miles! The exclamation mark is placed at the end of the last sentence, because, as a father of two girls (now grown women with their own children), I consider Elizabeth’s role of chaperone for that long trip to be no less than heroic. The trip was memorable because they found an isolated key that was good for camping, swimming, snorkeling, canoeing in the Everglades and all the good things that young people can find to do on a holiday.

Elizabeth’s alfresco life with her children perhaps reflected her own early childhood in Anoka, Minnesota. Recall that growing up in the small Minnesota town, Elizabeth enjoyed skiing and loved ice-skating. When the Minnesota winter waned and the days grew longer and warmer, she played tennis and croquet determinedly (but not expertly). Obviously Elizabeth took pleasure in the fresh air of the out of doors during her early years. It is not surprising then, to find that when she had her own children, she resumed enjoyment of outdoor life with them.

Was Elizabeth a Feminist?

Elizabeth Cavert Miller's life spanned a period of social upheaval especially as it pertained to women. Elizabeth was born May 2, 1920. The 19th Amendment to the Constitution of the United States granting women the right to vote was ratified on August 26, 1920 just a bit over three months after her birth.[11] Perhaps of greater importance to Elizabeth's life was the post-World War II struggle of women to obtain the same rights, privileges, and opportunities as men. This movement had its origin around 1950 and peaked around 1970. These were the years that Elizabeth's career was gaining national recognition. Unquestionably she was well aware of the revolution of the convergence of women's rights with human rights.

So, it is appropriate to ask whether Elizabeth was a feminist? Perhaps that is an odd question to place in a chapter concerning the Miller family life and the relationship of a husband and his wife. It seems to be appropriate to consider the issue here considering the closeness of Elizabeth and James. However, a problem for this author is that I have found no document of any kind that addresses the issue of Elizabeth's feminism. Her daughters give a conditional negative answer to the question.

Feminism, as certainly most people know, relates to women's rights in politics, society and economic opportunity. It often takes the form of militant expression of rights. According to her daughters, while she was not an activist, Elizabeth did *vigorously* support women's rights, including abortion rights. She believed that women had the right to, and needed, the same access to education and job opportunities as men. Further, she believed success was earned by hard work—and this applied to success in personal endeavors as well as to professional achievement.[12] Elizabeth, however was not politically active and never participated in a public demonstration. It is my view that Elizabeth was a feminist, but she was not one to wear her feminist beliefs on her sleeve—she was restrained in vocalization but un-muted in her personal endeavors and

11 The right of women to vote had local appearances earlier. Wyoming is an example where women gained the right to vote in all elections in 1869—the first state of the United States to do so.

12 Betty's attitudes toward success is reminiscent of John Houseman's (1902–1988) role as a curmudgeonly professor in the CBS television series "The Paper Chase". Houseman reprised the role in a series of television advertisements for the brokerage firm of Smith Barney, where he stated that "We make money the old fashioned way. We *earn* it." Betty achieved scientific recognition the old fashioned way, "she *earned* it."

professional achievement. It is not surprising, considering the closeness of the couple, that Jim shared Elizabeth's views on women's rights.

The few non-scientific organizations to which Elizabeth belonged were not those that are ordinarily associated with rampant feminism. Examples include the P.E.O.[13] Sisterhood. Elizabeth's mother and sister were both members and perhaps Elizabeth's affiliation was in part to please her mother. Elizabeth was a co-leader of a Girl Scout troop/group during the time that her daughters were active. Mentioned earlier in this chapter is the First Unitarian Church of Madison. The author knows of no other "organizations" to which Elizabeth belonged. While she seemed not to have been a member of Wisconsin's National Organization for Women (NOW), her portrait did indeed adorn the June 1986 calendar of that organization (Figure 8.2).

FIGURE 8.2 *Wisconsin Chapter of the National Organization for Women (NOW) calendar on wall of library cubicle. The image of Elizabeth Cavert Miller is clearly displayed on the month shown; June 1986. Calendar courtesy Karen Godshaw.*
PHOTOGRAPH BY THE AUTHOR.

13 P.E.O. (Philanthropic Educational Organization) was founded in 1869 as a pioneer woman's association in the United States and Canada. P.E.O. operates Cottey College, Nevada, Missouri. The college was founded by Virginia Alice Cottey Stockard, the great aunt of Emeritus Professor Arthur Campbell, Saint Olaf College, Northfield, Minnesota. Professor Campbell is the other grandfather of my two grandsons.

Whilst not a feminist in the usual meaning of the word, Elizabeth was a world-class pioneer cancer researcher—that much has been made manifest by her scientific publications and her numerous awards. Her scientific creativity and productivity removed barricades and smoothed the pavement for women who followed.

James A. Miller from His Daughters' View

What are Linda's and Helen's memories of their extraordinarily distinguished father? Elizabeth played the traditional role of the maternal parent with all that entails. James Alexander Miller was a masculine version of a traditional parent of the time. He was a loving and affectionate person but also a bit more distant than their mother. While he was less involved in helping with the girl's homework, he was nearly always at home in the evenings and available when needed. His expectations for the girl's school work were high. As Linda phrased it, when she got an A in a class, her father would ask "Why not an A+?" The question clearly was not to intimidate—it was asked humorously; nevertheless, it was intended to be taken seriously. In other words, why *not* an A+?

Jim had a good sense of humor. In contrast to the somewhat placid Elizabeth, he was much more volatile and emotional, but he would calm down quickly if he became upset. Outside of home and the lab, Jim was not a very social person. Differences between Elizabeth and Jim were apparent in their letter writing as the girls became older. Elizabeth would write frequent short notes and call weekly to arrange visits. Jim's letters to his daughters were long and often contained comments concerning the news of the day or he would reminisce about past happenings. Jim had coronary artery bypass surgery[14] and prior to the operation, he wrote warmly affectionate letters telling Linda and Helen how much he loved them.

Jim was fond of 19th and early 20th century instrumental music. He had a fine record player and an extensive record collection which he used on the occasion of Sunday noon dinner at home. Sunday dinner was the one "formal meal" of the week and Jim enjoyed selecting the music for that day to be enjoyed by the whole family together. Jim also enjoyed going to concerts. Despite Helen's modest competence at the piano, Jim enjoyed sitting in a chair and listen to his young daughter practice. Helen's playing was a source of pride and pleasure.

14 The surgery is also known as coronary artery bypass graft or, more simply, "CABG". Jim's surgery was during the early 1980s well before his death in 2000.

The Miller home at 5517 Hammersley Road in Madison, Wisconsin was full of books. Helen observed that some people use coffee tables for their named function, i.e., a place to set a coffee pot and some cups and saucers. Not so in the Miller household. Coffee tables were for stacks of books, and overflowing book shelves were to be found throughout the house. Dad was the real reader and it was he who would buy and read non-fiction books. He read the journal *Science* at home as well as the *Wall Street Journal*, the *Milwaukee Journal*, and *The Capital Times* on weekdays and the *Wisconsin State Journal* on Sundays.

Jim liked to work in wood and it was he who helped Helen with her first lesson in carpentry. Helen built a box with her dad's supervision. The box had a hinged top suitable for the storage of the "treasures" of a young girl. In itself, building a box does not sound like much, but Helen remembered it—probably because of the attention that was paid to her by her father. Of course, a man who works in wood does more than teach his child how to build a box. Jim built the marvelous playhouse that the girls had in their back yard. Fond memories of that playhouse remain with the daughters. Jim built two Adirondack chairs for the backyard of the family home. And, his carpentry efforts did not end with home projects. Jim constructed test tube and beaker racks and other small items for the laboratory—these were not complex construction projects but they were necessary in a laboratory at the time.

It was the father's role to maintain the family automobile and Jim assumed the responsibility. Jim had a high level of safety awareness and had the family's blue Buick outfitted with passenger safety belts long before that was common. He also added red reflection tape to the back bumpers of the Buick and the tape became a marker such that the girls' friends could always recognize the rear of the Miller Buick at a distance.

The job of recording family events on film (it was film in the days prior to contemporary digital cameras) was Jim's and he took many color slides of his closely-knit family. Jim is *not* in many of the family photos because he was the camera operator. Cameras were not "automatic" in the days when the Miller children were young and the operation of a good camera required some technical skill. Jim liked that and this is probably related to his love of gadgets. Jim was fascinated by his stereo system, cameras, movie projectors, and even fish locators.

Jim built two telescopes to add to those that he bought. This was a continuation of the telescope fashioning skills he developed as a young man. Unfortunately, the lights of Wisconsin's capital city precluded clear viewing, and Jim had to take his telescopes out of the metro area for proper viewing.

Jim Miller watched the evening news with "the most trusted man in America", CBS's Walter Cronkite. It was the same Cronkite who early on narrated the launching of space flights, the assassination of John Fitzgerald Kennedy and other notable events of the time which Jim watched carefully. On Friday evenings while Mom ironed clothes, both she and Jim watched National Public Television's *Washington Week in Review*. There were other television programs that Jim enjoyed, *Nova* and *Masterpiece Theater* to mention but two. It was TV watching time as a family on Sunday evenings when all enjoyed "*The Wonderful World of Disney*". Linda has suggested that her father probably watched Disney with some reluctance. Jim also watched the famous National Football League's Green Bay Packers perform on television. While Jim did not ordinarily drink alcoholic beverages, he did partake of a beer or two while watching the Packers, and he let it be known to anyone who was nearby whom he rooted for—the Packers, of course. Earlier, it was recorded that Jim helped Helen with wood-working. While Jim was competent with hammer and saw, for leisure he preferred the Packers (and reading books and watching specials on television). Jim also watched the University of Wisconsin's football Badgers.

Family Gatherings in Madison

Elizabeth's sister, Jane Cavert Mitchell and her husband John E. Mitchell, better known as "Jack", also lived in Madison. The Mitchells' had three sons, William, Charles, and Robert, better known as Bill, Chuck, and Bob, and the Millers often got together at Thanksgiving and Christmas times. The two families had much in common. Both fathers, Jim Miller and Jack Mitchell, had Ph.D.s and both had significant positions at the University of Wisconsin, Madison. Of course their spouses, Jane and Elizabeth, were sisters. Jack grew up in the same part of Saint Paul, as did Jane and Elizabeth. Jack's undergraduate degree was in Agricultural Biochemistry, as was Elizabeth's. The cousins and their aunts and uncles had great times together during the holidays.

Trips to Minnesota

Elizabeth was born is Minnesota and graduated from high school and undergraduate college in Minnesota. Elizabeth's parents still lived in Minnesota when Linda and Helen were children. Consequently, August visits to Elizabeth's parents' home in Saint Paul became a family tradition which permitted Linda and Helen to know their grandparents and the grandparents to keep up with the growing granddaughters. It would be strange indeed if Elizabeth did not enjoy going "back home" to enjoy the company of her parents and other family members as well as visiting familiar places of her youth.

It is not a difficult drive from Madison, Wisconsin to Saint Paul and Minneapolis, Minnesota and the Miller family made the traditional driving trip with Jim at the wheel a number of times. Perhaps a more fun trip for Linda and Helen was to take the train to Portage, Wisconsin north of Madison and then take the westbound Milwaukee[15] Road's famous Hiawatha train that started its westward trek in Chicago.[16]

The Miller family stayed with Elizabeth's parents at the Cavert family home on Grantham Street, Saint Paul. What did the girls do when they visited the grandparents? Well, for one thing, Helen climbed the small apple tree in front of the house. Helen and Linda also enjoyed visiting their mom's brother and his spouse, Uncle Mead and Aunt June, who lived nearby in Minneapolis. The Mead Cavert lawn is where the famous croquet games were played. I have no record of who played best, but it is legendary that the girls' grandfather, Will Cavert, played with a mean mallet and he did not give his grandchildren any leeway. Well, no leeway except when he "partnered" with one of the younger grandchildren.

The Minnesota State Fairgrounds, contiguous to the University of Minnesota Saint Paul campus and only a few blocks from the grandparents' home, was visited when the Miller family was in Saint Paul at fair time. The Minnesota State Fair is no inconsequential attraction in Minnesota. It attracts more attendees daily than any other comparable fair in the United States.[17] The fair boasts displays of horses, beef cattle, dairy cattle, swine, sheep, goats, chickens and turkeys, to mention only some of the agricultural animals to be seen. Artists, both pictorial as well as culinary, compete for blue ribbons. All of that is free with the price of admission—and the list also included Machinery Hill where goliath farm machines were on display for kids of all ages. Truly, the Minnesota State Fair is a wondrous site for young and old alike.

Grandpa Cavert enjoyed taking Linda and Helen to the barns where he could expound on the breeds of cows and other beasts. Showing his granddaughters the State Fair was as much fun, if not more so, as taking his own children to the fair a generation earlier. Grandma took the girls to the Woman's Building so that Linda could see the needlework on display. Lunch was obtained at one of several church stands with home-cooked victuals. The girls must have been

15 Corporate name: Chicago, Milwaukee, St. Paul & Pacific Railroad.

16 The Hiawatha had its origin in the mid-1930s and was pulled by one of the fastest steam locomotives ever built that roared from Chicago to Minneapolis/Saint Paul at speeds in excess of 100 miles per hour.

17 The Texas State Fair has a greater total number of visitors but it lasts twice as long as the Minnesota fair. Nearly two million visitors annually were attending the Minnesota fair in the first decade of the 21st century.

exhausted by the time to return home—exhausted, but probably looking forward to the State Fair of the next year.

Linda and Helen visited the Gibbs Farmhouse Museum of Pioneer and Dakota Life with its 1849 house, two barns, and a one-room school house. The Gibbs Farm was conveniently located only a few blocks north of the Cavert grandparents' home. Probably the grandfather, William Cavert, had a bit of extra pleasure in showing the farm museum to his granddaughters—he served on the board of the Gibbs. Another favorite place to visit was the American Swedish Institute in Minneapolis. The Institute is housed in a 33 room castle-like mansion (it is probably more accurate to call the architecture "chateauesque", but many young people think it looks like a castle) with its museum and displays. Another place to visit, while in Minnesota, was the botanical gardens of the University of Minnesota, better known as the Minnesota Landscape Arboretum. The gardens are home to 5,000 kinds of plants beautifully arranged in over 1000 acres. When she was older, Helen's interest in the botanic garden was enhanced by the knowledge that her grandfather was involved in the Arboretum's early days and her cousin, Harlan Cavert (Mead and June Cavert's son), now works there as a development officer for the Arboretum. Probably the favorite activity of Helen when visiting Saint Paul was visiting Como Park where her mother would rent a canoe and the two enthusiasts would paddle happily on beautiful Como Lake.

I have been a resident of the twin cities of Minneapolis and Saint Paul off and on since 1953. Because of this, I instantly knew what Helen was talking about when she told me of the fascination of going to "downtown" Minneapolis for back-to-school shopping and lunch at Dayton's Department Store. Helen described Dayton's as an "amazingly big store and quite a treat to visit". I remember Dayton's as elegance, variety, abundance and escalators; floor after floor of the most beautiful consumer goods that I had ever seen, and that includes department stores in New York City—and Paris and Tokyo, for that matter. No wonder it was a treat for the girls from Madison. No big-box discount store of today can compete with the absolute sophistication and refinement of the downtown Dayton's of former years. Clearly, coming to visit the Cavert grandparents was more than just sitting in their living room or climbing a tree in the front yard.

Trips to Other Places

James and Elizabeth Miller attended many scientific meetings throughout the United States and foreign countries. Sometimes their daughters went with them. Helen remembers when her mother and father showed them the sights of Washington, D.C. Helen was obviously young. Instead of focusing on the classic architecture of the National Archives or the Capitol as her mother may have wished, Helen fondly remembers the pigeons.

Jim and Elizabeth were invited to participate in The Ninth International Cancer Congress, held in Tokyo, October 23 to October 29, 1966.[18] Helen was in the 6th grade and Linda was in the 9th grade. Helen said that her Mom and Dad did not spend the evenings drinking at the bar and politicking with other scientists in Tokyo. Rather, they were more comfortable with their daughters during the evening hours. Obviously, the scientific sessions of the meeting held little interest for Linda and Helen. The organizers of the meeting had a number of "spouse" tours and the girls were welcomed to tag along and probably saw more of Tokyo than did either of their parents. Helen remembered that the Japanese hotel where they were staying had Western meals. She ordered spaghetti night after night. Why? Well, the pasta was luscious but she especially was delighted by the fancy silver container the sauce came in. Permission to order spaghetti every night reveals how gentle and benign Jim and Elizabeth were to their younger daughter.

James Miller loved to read about Charles Darwin. At Jim's retirement in 1985, his graduate students and post-doctoral associates gave him and Elizabeth a trip to the Galapagos Archipelago. As all biologists know, the Galapagos were visited by Charles Darwin, who spent from September 16 to October 18, 1835, exploring the islands. Darwin was billeted aboard the HMS Beagle in its circumnavigation of the world, and the unique fauna of each of the Galapagos islands played a crucial role in the formulation of Darwin's concept of evolution. How could you be a biologist (in Jim and Elizabeth's case, biochemists) and not be thrilled about going to the Galapagos? Miller, age 70, at the time, was indeed thrilled, and he took, along with his wife, members of the family on this trip of a life-time. Helen and her husband Dave went on the memorable trip. Linda and Russ were invited but were unable to go at that particular time,

18 James and Elizabeth Cavert Miller returned to Tokyo for the 9th International Symposium of the Princess Takamatsu Cancer Research Fund in 1979. The Proceedings of the symposium *Naturally Occurring Carcinogens-Mutagens and Modulators of Carcinogenesis*, Tokyo: Japan Scientific Societies Press, were edited by E.C. Miller, J.A. Miller, I. Hirono, T. Sugimura, S. Takayama. The participants of the symposium are illustrated in the frontispiece of the volume with Elizabeth seated next to His Imperial Highness Prince Takamatsu and James seated next to Her Imperial Highness Princess Takamatsu. The frontispiece is a full color image clearly showing how the Millers looked in 1979. For history buffs, H.I.H. Prince Takamatsu was a brother of Japan's Emperor during World War II, Emperor Hirohito (now known in Japan as Emperor Shōwa).

so Elizabeth's sister and brother-in-law Jane and Jack Mitchell went in place of Linda and Russ.

Elizabeth was as excited as Jim to visit the islands. She was fascinated by invertebrate "critters" along the sea shore. She must have been a competent swimmer because she snorkeled to see even more. Biologist son-in-law David Alexander (Helen's husband) was crucial in the identification of the marine creatures. I asked Helen if the family got to see the giant tortoises[19]—the answer was "yes". And did they see the giant lizards (marine iguanas) that spew salt from their nostrils—again "yes". Obviously, it was a fun and instructive holiday for a family.

The good times did not stop with the visit to the islands. Next on the itinerary was a trip to the majestic snow-capped mountains and active volcanoes of Ecuador.[20] Quito, the capital, is at 9,000 feet elevation. The Millers viewed these highlands on a train with open windows and a very permissive trainman, who allowed passengers to ride on the top of the railroad car clinging to the top walkway (cat-walk). Elizabeth Miller was not to miss this kind of ride. Elizabeth, Helen and David, as well as other tourists, rode up high on the top of the railroad car tightly clutching the walkway, but none were of the age of Elizabeth. Further, her loyal and loving husband elected *not* to follow her to the roof. It is not recorded, but I can almost hear Jim saying (at least to himself) "Lizzie, get down; come down, Lizzie."

James Miller did not come from a snobbish or affluent family, and there is evidence that he never changed the ways of his modest origins—including the times that he took his meals. Consider the scientific meeting the Miller family attended on the island of Antigua[21] in the Caribbean. The island is small, and while independent, it was once under British control. The timing of meals was distinctly European. Jim liked his supper early and he rankled over the long wait for the evening meal until after 7 p.m. During the scientific sessions at the Antigua meeting, Linda and Helen made good use of the beach and played and swam—delightful perks for the offspring of scientists attending a serious scientific meeting.

---

19 Galápego is the Spanish masculine noun meaning turtle. It is lore that Spanish sailors named the islands after the giant turtles (tortoises).

20 The Galapagos Islands are located on the equator. So too is the country of Ecuador from whence its name comes. The capital is Quito which is located 0° 13′ S (or, about 15 US Miles south of the equator).

21 Christopher Columbus named the island for the church in Seville, Spain known as "Santa Maria La Antigua" (Saint Mary the Old).

Higher Education for Linda and Helen

Linda went to the University of Wisconsin, Madison for an undergraduate degree in textile design, which was followed by a masters degree from the University of Wisconsin-Stout, located in Menomonie, Wisconsin. Linda met Russ Forbess in the 10th grade. They were married on July 20, 1974 in Madison, Wisconsin.

Helen is a biologist. Where did her biological interests have their origin? Both of her parents had outstanding teachers during their high school years, which may have contributed to the parents' sustained commitment to science. Not unlike her parents, Helen had an exceptional high school biology teacher by the name of LeRoy Lee.[22] Mr. Lee not only was a good teacher in the classroom, but he took his students on a week long trip to Arizona to study the ecology of southwestern deserts.

Helen took her undergraduate degree at the University of Wisconsin-Madison but she did not live at home. She lived in a dormitory for two years followed by an apartment for another two years. Of course she came home on some weekends; probably for nourishment and laundry. As with so many young people who are finishing their undergraduate years, Helen did not know for certain what she wanted to do. However, knowing that she liked science, Helen applied to Duke University in Durham, North Carolina, where she was admitted to a Ph.D. program. Elizabeth Cavert Miller drove her daughter the more than 900 miles to Durham to help Helen get settled in her new academic home.

Helen called home from time to time while she was in graduate school. It almost seemed like she had a second major professor when her dad sought news of how her research was going. Mom seemed to understand that Helen did not need the stress of reporting on research during a telephone conversation. The research must have gone satisfactorily, because several years later, Helen obtained her Ph.D. in 1982 from Duke University. Her husband, David Earl Alexander, also received his Ph.D. from Duke. Helen and David were married on October 10, 1981 in Raleigh, North Carolina. David, as is Helen, is also a biologist at the University of Kansas. Prior to Kansas, Helen was appointed for a year and a half to a postdoctoral fellowship at the Cereal Rust Laboratory, now known as the Cereal Disease Laboratory, on the Saint Paul campus of

22 Helen Alexander's high school biology teacher, LeRoy R. Lee, was designated a Distinguished Alumnus by his undergraduate alma mater, the University of Wisconsin River Falls (B.S., 1957). Lee obtained his master's degree at the University of Michigan in 1963. He received many teaching honors and served as president of the National Science Teachers Association.

the University of Minnesota, where she was not far from her grandparents' home on Grantham Street. Helen's husband David took advanced engineering classes at that time. The Alexanders moved to Louisville, Kentucky in 1984 where David had accepted a position as an assistant professor at Bellarmine College.[23] Helen and David remained in Kentucky until December 1987, when the couple moved to Lawrence, Kansas where they now reside.

An Observant Grandmother

Childhood memories are wonderful. So too are recollections from adult years. A touching memory is that of the adult Helen, with her baby daughter, Diane visiting Helen's mother in Madison. Baby Diane seemed well enough to her new mother. Grandmother Elizabeth was not so certain. She observed that her granddaughter Diane was not as active as she thought she should be. Elizabeth asked her daughter Helen to take Diane's temperature. It was 105° F. What to do? Grandmother simply said, "Get to the Emergency Room at the hospital; they will know what to do." Baby Diane had septicemia ("blood poisoning") with an unknown cause. The young child spent the next 10 days in the hospital. The happy ending to this story is that there were no untoward after effects. Diane came home well and happy. However, the outcome could have been much more dire were it not for an observant grandmother, who herself was seriously ill at the time. Helen was lucky indeed to have such a caring mother—and Diane was lucky to have Elizabeth as a grandmother.

Elizabeth's Terminal Illness

Elizabeth suffered from respiratory problems in the fall of 1986. Scans (computed axial tomography, CAT) of her upper body revealed an abnormal density in her lungs. Surgery was performed to remove the affected lung tissue with the result that she could breathe better. However, even after surgery, Elizabeth had a constant feeling of low energy and she seemed to be in poor health. Subsequently a full body scan revealed kidney cancer. A kidney was removed that was judged to be metastatic. Elizabeth received radiation after the kidney surgery, which proved to be exhausting to her. Ultimately, her weakened condition led to her being confined to a wheelchair. Jim Miller had health problems

---

23 The college is now known as Bellarmine University, a Catholic liberal arts institution, with more than 2,300 students. It is the largest private university in Kentucky.

that included a heart attack. Accordingly, Elizabeth was considered by Jim to be the healthy one. It was a shock for Jim to learn of the seriousness of his wife's illness.

Linda, who lived in Schofield, Wisconsin, less than 150 miles from Madison, spent much time with her mother during that last illness. Helen, who lived in Louisville, Kentucky at the time, was unable to spend as much time with her ill mother. Nevertheless, Helen was able to come to Madison monthly and spend a few days each time. Because Elizabeth was aware of the prognosis of her disease, she sought retirement. This was granted and she received the title of Professor Emeritus on June 1, 1987. By the autumn of 1987 she was acutely ill. It was then that Linda was especially helpful. She got the "right people" to help at home but before long, Elizabeth was admitted to the hospital in great pain.

There is no record of what Elizabeth thought when, during her hospitalization, when Jim was hit by a car. He announced to Linda and Helen from the hospital that "now, we are both here." Jim had been crossing a street to visit Elizabeth in the Meriter Hospital of Madison. The accident resulted in a broken wrist. Jim was hospitalized for only a short period of time but long enough for the surgical insertion of pins to stabilized his wrist.

Elizabeth's mind remained active and aware despite her discomfort. For example, Helen recalls her Mom, late in her illness, thoroughly enjoying a public TV nature show on pelicans. However, Elizabeth was never a person to while away the hours without substantive work and this proved true even during her final stay in the hospital. She made it known to her students that they were welcome to confer with her about their current research and the implications of this for future experiments. Students visited her bedside at the hospital until a few weeks prior to her death. Elizabeth Cavert Miller died on October 14, 1987 at the age of 67 in the University Hospital (University of Wisconsin Hospital and Clinics). A memorial service was held for her at the University of Wisconsin Pyle Center on October 25, 1987.

Barbara Butler Miller

Jim Miller married Barbara Butler on 21 December 1988.[24] It is appropriate to end this chapter with that marriage and the years Barbara and Jim spent together as a married couple. What Jim did in retirement tells much about him, and that information was obviously best obtained from Barbara. Jim died on

24 Much of the material contained in this section was obtained in an interview with Barbara Butler Miller in her home in Madison, Wisconsin on 17 November 2010.

24 December 2000. The 12 years of their marriage were happy ones and were described by Barbara as "very fine".

Barbara was no stranger to the Millers prior to her marriage to Jim. Beginning in 1965, she worked for nine years as a technician/laboratory specialist in the Miller laboratory at the McArdle in Madison, Wisconsin. Barbara came on board with an undergraduate degree in zoology and minors in chemistry and bacteriology earned in 1964 from Miami University[25] located in Oxford, Ohio. She was by training well prepared to work on the biochemistry of chemical carcinogenesis. After her tenure at the Miller laboratory, she earned a Master's degree from the University of Detroit and her Ph.D. from the University of Michigan, Ann Arbor. Those advanced degrees will be discussed later in this section.

I asked Barbara how she felt upon entering the laboratory of the distinguished Millers in the early days of her employment. Barbara, while fresh out of college, was not in any way intimidated by the Millers. Neither Elizabeth nor Jim acted as if they were famous. Jim was self-effacing and modest. Jim was pleased with the work that he and Elizabeth had already accomplished, but instead of basking in the glory of these past endeavors Jim was always looking forward to the next series of experiments.

Barbara appreciated how much she learned about the scientific method and the construction of well controlled experiments. Years later, Barbara still uses the skills she acquired in the Miller laboratory in aspects of her daily life and in her work.

According to Barbara, Jim and Elizabeth treated everyone in the lab equally and this was especially so with regard to graduate students from abroad. The Millers would generously provide funds for students to visit home and to attend scientific meetings. Generosity was not just with scientific affairs but extended to informal social activities such as laboratory picnics, at which the Millers provided all of the necessary picnic fare. These happenings, greatly appreciated by hungry graduate students, were often held at the Stewart County Park near Mount Horeb, Wisconsin only a few miles west of Madison. Barbara stated that everyone in the laboratory had fun and greatly enjoyed these casual and unpretentious times together. Through the several years that she worked in the laboratory, Barbara came to know the Miller daughters and by her own

25 Miami University first opened its doors to students in 1823. It is a state supported University that is known for its commitment to personalized education for undergraduate students. It is named for the Miami Indian Tribe that lived in the area at the time of its founding. It has been said that the poet Robert Frost deemed Miami as "the most beautiful campus that ever there was."

words, became "very attached" to them. It was clear that the memories of her McArdle Laboratory days were cordial and warm, and her feelings might even be said to be familial.[26]

After completing nine years in the Miller laboratory, Barbara moved to Detroit, Michigan. While there she changed her career path from science to the humanities. In 1979 she received an M.A. in Religious Studies from the University of Detroit.[27] As an adjunct faculty member she taught Biblical Studies at the University of Detroit from 1981 to 1989. In 1984, she began Ph.D. work in Biblical Studies in the Near Eastern Studies Department at the University of Michigan, Ann Arbor. It was a difficult time financially commuting and teaching. Another former Research Specialist from the Miller laboratory shared her dilemma with the Millers who immediately sent financial assistance. Barbara said that she had never experienced such generosity in her life.

Because of the close relationship of Barbara with the Millers, it is not surprising that Barbara visited Madison when Elizabeth was terminally ill. Of course, Barbara recognized the seriousness of Elizabeth's illness, in part because she had worked in a cancer laboratory for nine years. Members of the family including the two daughters were similarly aware of Elizabeth's grave condition. Because the Millers were very reserved people who "kept their private lives pretty much to themselves", few others knew of Elizabeth's perilous illness.

Not long after Elizabeth's death, Jim expressed an interest in Barbara. Barbara was still taking courses at the University of Michigan, and because of her course work and her judgment that Jim should live alone for a time to come to terms with his grief, she demurred. Jim waited but pressed his case. Barbara relented, and the couple was married December 21, 1988.

Jim already had a home on Hammersley Road in Madison, and that seemed a reasonable place for the new couple to live. Barbara was a gardener and planted a sizable vegetable garden and small perennial beds in the ample yard. Barbara continued working on her Ph.D. with the enthusiastically "wonderful" support of Jim. She was awarded the degree in 1994 from the University of Michigan, Ann Arbor.

---

26 Non-academic readers may be surprised by the use of "familial". A colleague of mine used to say that we were not a department; instead, he affirmed that we were a "family". Scientists working together for common goals—if circumstances are right—may indeed become almost like a family.

27 The University of Detroit, founded in 1877, merged with Mercy College founded in 1941, to become the largest Catholic University in Michigan. The merger was in 1990 and the institution is now known as the University of Detroit Mercy. The combined institution is sponsored by the Jesuits and Religious Sisters of Mercy.

Because of her prior teaching experience and her quest for a Ph.D., Barbara was recruited to teach at Edgewood College in Madison, Wisconsin. Edgewood is a small liberal arts college of 2,400. It is sponsored by the Sinsinawa Dominican Order of the Roman Catholic Church located in Sinsinawa, Michigan. Barbara taught at Edgewood until she retired as Professor of Religious Studies in 2006.

Jim and Elizabeth had spent wonderful afternoons together watching the famous National Football League's Green Bay Packers. Jim's appreciation for the Packers has been mentioned earlier in this chapter and his loyalty to the team diminished in no way in his later years. Barbara's appreciation matched Jim's and they both cheered when Green Bay players performed as they should. Jim was alleged to have enjoyed good beer, and perhaps a pint or two was consumed during the Green Bay games.

A somewhat more serious form of entertainment was attendance at the Madison Symphony Orchestra, a leading regional orchestra. The concerts were held in the splendid Overture Center for the Arts in downtown Madison just a bit more than a block from the Wisconsin State Capitol. Barbara and Jim enjoyed going alone or with son-in-law Russ Forbess' parents. Outstanding classical music in a beautiful venue leaves memories that never fade. But it was not all serious music that the couple enjoyed. Jim and Barbara delighted in the comic operas of W.S. Gilbert and Arthur Sullivan which were put on several times of year.

During the happy days that Jim and Barbara spent together, Jim continued his longtime habit of selective television watching. As mentioned previously, he loved to watch the Green Bay Packers regardless of whom they were playing. He was absolutely faithful to Public Broadcasting Service's (PBS) The MacNeil/Lehrer News Hour, but this he watched alone. Barbara had to decide when she could work on her Ph.D. courses, and she used the News Hour for a portion of her study time. Obviously, she had other study times as well. As stated previously, she received that highest academic degree in 1994. Her dissertation was titled: *Women, Death, and Mourning in the Ancient Eastern Mediterranean World.*

Barbara and Jim travelled extensively together. Jim's daughter Helen Alexander was invited to give a scientific talk in Japan. What better time to travel to Japan and enjoy sightseeing, with the pleasure of meeting with former post-doctoral colleagues of Jim. Barbara and Jim visited Hiroshima[28] among

28 Your author made several trips to Hiroshima in 1994 in regard to the organization of a scientific meeting. It was amazing for me to see a robust, vigorous, bustling, thriving modern city that replaced the devastated Hiroshima of World War II. I suspect that Barbara and Jim were equally amazed.

many other notable places in Japan. Another trip ensued from Jim being invited to lecture in Taiwan. Barbara described that, as in earlier trips, Jim met with former students. There was much pleasure with the exchange of gifts and memories of shared times at the Madison laboratory. After Taiwan, it was off to Korea for another series of lectures and the great pleasure of meeting with former students.

Not all of the trips were to give lectures and meet with former colleagues. A trip of great pleasure was to England with Linda and Russ Forbess (Jim's daughter and son-in-law) and Barbara's daughter Sarah. I know not if Jim and Barbara were ardent Anglophiles but Barbara described the trip as replete with pleasure and that they had a "wonderful time".

Perhaps they had even more pleasure in Hawaii. The reason for this statement is that the couple visited the Pacific islands three times. "Everybody went" in 1994 and "everybody" included Barbara's daughter Sarah and her fiancée, Helen and Dave and their two children Diane and Kevin, Linda and Russ Forbess and, of course, Barbara and Jim. The family entourage walked the beautiful Waikiki Beach on Oahu on Christmas day—the turquoise waters of the Pacific Ocean enhanced the already beautiful beach providing for "wonderful" life-time memories. However, not all trips were to a tropical paradise.

Barbara and Jim traveled to Alaska with Russ and Linda Forbess, and this was followed by a trip to Door County, Wisconsin. Door County is located on a peninsula jutting out to the northeast of Green Bay about 150 miles from Madison. Door County is surrounded on three sides by Lake Michigan, and this piece of land has become a favorite destination for many people. Barbara and Jim rented a house large enough for the same family group that had gone to Hawaii in 1994. The family group could not help but enjoy the serenity of Door County.

When not traveling, Jim enjoyed his many interests that varied from reading about Darwin to astronomy (a love from his boyhood that continued into his later years).

Jim's Last Illnesses

James A. Miller had two homes that he obviously loved. One was the home on Hammersley Road where he lived with Elizabeth and his daughters and later with Barbara. The other home was his laboratory. Barbara put it this way: The "lab was his life". As people age, accommodations for the passing of years must frequently be made. Such was the case with Jim.

In 1998 Jim had a fall in his garage. The fall was a portent of things to come. Jim fell again, on ice, going from the parking lot to his laboratory at the McArdle.

This time the consequences were far more serious. Jim suffered a compression fracture of his spine. A compression fracture, a condition wherein a vertebra collapses, may result from trauma, osteoporosis, or cancer. A retrospective speculation of why Jim had the compression fracture is guess-work at best. However, trauma is one cause of compression fracture, and Jim obviously fell with traumatic consequences. Osteoporosis may have been a contributing factor to the compression fracture. Jim, born in 1915, was not young at the time, and osteoporosis is far from uncommon in older people. And finally, Jim had prostate cancer. Metastasis of prostate cancer to bones, including the vertebrae, is not uncommon. A major symptom of compression fracture is pain. Jim suffered.

Unfortunately, Jim was also afflicted with diabetes, and this condition included diabetic neuropathy. In some cases, peripheral neuropathy can result in loss of balance and Jim was not steady on his feet. Jim also suffered from congestive heart failure. Considering the aggregate of his problems, one cannot but wonder about the drive and determination that kept Jim going in his later years. Jim was hospitalized after his fall on ice outside his laboratory at the McArdle.

After a short time in the hospital in 1998, it was considered best for Jim to be moved to the Middleton Village Nursing and Rehabilitation Center.[29] The nursing home finished Jim's life in the lab. It did not, however, finish his interest in life, and in this case, Barbara was his greatest asset. Barbara arranged for Jim's room in the nursing home to appear "homey" with framed pictures of Jim's mother, his aunt and his daughters. Jim wanted his Ph.D. diploma hanging on the wall and Barbara saw to it that it was.[30] She made arrangements for the room to have a desk and typewriter.

Barbara considered moving Jim back home, which Jim would have dearly appreciated. She had the Hammersley Road home "analyzed" but it was found inadequate for wheelchair accessibility. During the time in the nursing home, Barbara loyally took Jim home almost every Sunday until the physical pain became too great.

Jim died on December 24, 2000 at age 85. Memorial services were held on January 14, 2001 at the Pyle Center close to the shore of Lake Mendota. There is a grave marker in the Forest Hill Cemetery on Speedway Road, Madison, Wisconsin, engraved with the names Elizabeth, James, and Barbara. To honor

29 The Middleton Village Nursing and Rehabilitation Center is located on Elmwood Avenue in Middleton, Wisconsin. Middleton is a city contiguous with Madison, Wisconsin.

30 Your author noted the contrast between the diploma on the wall and the multitude of award plaques consigned to a closet at home. I think it shows that Jim had his priorities in order.

Jim, the McArdle Laboratory established the James A. Miller Fellowship Fund to assist graduate students in the study of cancer.

As the art of an artist remains alive after the departure of the artist,[31] so too do the scientific contributions of Elizabeth and Jim (Figure 8.3). Perhaps even more importantly, their work lives on through that of their many students.

FIGURE 8.3 *Elizabeth C. Miller and James A. Miller in 1985.*
COURTESY OF THE UNIVERSITY OF WISCONSIN-MADISON ARCHIVES, IMAGE #S05537.

31 When I gaze upon Sandro Botticelli's *Nascita di Venere* (the Birth of Venus) in Florence's Uffizi Gallery, I see an abundance of vitality in that painting which I suspect was there when it was painted in 1486. Venus's grace and allure did not fade.

APPENDIX 1

Awards, Honors and Professional Activities: Elizabeth Cavert Miller (ECM) and James Alexander Miller (JAM), Individually or Jointly*

L'homme qui devine l'énigma du cancer n'a pas besoin d'un Prix Nobel[1]
Attributed to the Belgian pathologist Dustin

1950–1955	Biochemistry Study Section, National Institutes of Health (JAM)
1954–1964	Assistant Editor, Scientific Editor, and Associate editor, *Cancer Research*, journal of the American Association for Cancer Research, Inc. (ECM)
1957–1960/ 1974–1977	Elected to Board of Directors, American Association for Cancer Research, Inc., President, 1976–1977 (ECM)
1960–1969	Food Protection Committee, National Research Council (JAM)
1961–1968	World Health Organization (WHO) Committee on Evaluation of Carcinogenicity of Food Additives (JAM)
1962	Ann Langer Cancer Research Foundation Award aka Langer-Teplitz Award for Cancer Research, Chicago Cancer Foundation (ECM & JAM)
1965	Lucy Wortham James Award for Cancer Research, sponsored by the New York Community Trust and awarded by The Society of Surgical Oncology (formerly known as the James Ewing Society) (ECM & JAM)
1965–1968	Board of Directors, American Association for Cancer Research, Inc. (JAM)
1967	National Cancer Institute Committee on the Evaluation of Cancer Research and Training in Chemical Carcinogenesis (JAM)
1968–1969	Committee on Evaluation of Safety of Non-nutritive Sweeteners, National Research Council (JAM)
1968–1972	Member, Pharmacology B Study Section, National Institutes of Health (ECM)
1968–1972	Research Advisory Council, American Cancer Society (JAM)

* Appendix 1 was taken, in part, from Conney, A.H., M.C. Poirier, Y.-J. Surh, F.F. Kadlubar. 2009. *Elizabeth Cavert Miller 1920–1987, James A Miller, 1915–2000. A Biographical Memoir.* Washington, D.C.: National Academy of Sciences.

1 English translation: "The person who discovers the riddle of cancer has no need for the Nobel Prize."

 | DOI 10.1163/9789004286801_010

1969	Program Committee, 10th International Cancer Congress (JAM)
1969	Advisory Panel on Carcinogenicity, Commission on Pesticides and their Relationship to Environmental Health, U.S. Department of Health, Education and Welfare (JAM)
1969	Panel on Non-Psychiatric Effects of Drugs of Abuse, National Institute of Mental Health (JAM)
1969	G.H.A. Clowes Award and Lectureship, American Association for Cancer Research (JAM)
1971–1972	Chair, Pharmacology B Study Section, National Institutes of Health (ECM)
1970–1973	Core Committee on Chemical Carcinogenesis, Smithsonian Institution (JAM)
1971	Bertner Foundation Award, M.D. Anderson Hospital and Tumor Institute (ECM & JAM)
1971–1975	Cancer Research Center Review Committee, National Cancer Institute (JAM)
1972–1975	Member, Scientific Advisory Board, National Center for Toxicological Research (ECM)
1972–1975	Colon Cancer Segment, Carcinogenesis Program, National Cancer Institute (JAM)
1972–1976	Council on Analysis and Projection, American Cancer Society (JAM)
1972–1976	Advisory Panel on Carcinogenesis, National Cancer Institute (JAM)
1973–1976	Member, Council for Research and Clinical Investigation Awards, American Cancer Society (ECM)
1975–1976	Chair, Council for Research and Clinical Investigation Awards, American Cancer Society (ECM)
1973	Wisconsin Divisional Award, American Cancer Society (ECM & JAM)
1973	Outstanding Achievement Award, University of Minnesota (ECM)
1975	Papanicolaou Award, Papanicolaou Institute, Miami (ECM & JAM)
1976	Rosenstiel Award for Basic Medical Research, Brandeis University (ECM & JAM)
1977	National Award in Basic Sciences, American Cancer Society (ECM & JAM)
1977	Review and Evaluation Committee on the Carcinogenicity of Chlordane and Heptachlor, National Research Council (JAM)
1978–1981	Board of Scientific Consultants, Memorial Sloan-Kettering Cancer Center (JAM)
1978	First Founder's Award, Chemical Industry Institute of Toxicology (ECM & JAM)

1978	Bristol-Myers Award in Cancer Research, Bristol-Myers (ECM & JAM)
1978	Election to the National Academy of Sciences (ECM & JAM)
1978	Gairdner Foundation Award, University of Toronto (ECM & JAM)
1978	Prix Griffuel, Association pour le Developpement de la Recherche sur le Cancer, Villejuif, France (ECM & JAM)
1978–1980	Member, U.S. President's Cancer Panel, National Cancer Institute, (ECM)
1979	Federated American Societies of Experimental Biology 3M Life Sciences Award (ECM & JAM)
1979	Louis and Bert Freedman Foundation Award in Biochemistry, New York Academy of Sciences (ECM & JAM)
1979–1983	Cancer Information Dissemination and Analysis Center (CIDAC)-Carcinogenesis Advisory Board, Franklin Research Center (JAM)
1980	Mott Award, General Motors Cancer Research Foundation (ECM & JAM)
1980	Honorary Member, Japanese Cancer Association (ECM & JAM)
1980	Medalist Award, Intra-Science Research Foundation (JAM)
1980	Honorary Member, Sigma Delta Epsilon Graduate Women in Science (ECM)
1980	Board of Directors, American Cancer Society, Inc. (ECM)
1980	Life Sciences Advisory Board, Los Alamos National Laboratory (JAM)
1980	First Annual Award Lecture, Laboratory of Nuclear Medicine and Radiation Biology, University of California, Los Angeles (JAM)
1980–1983	Scientific Board of Visitors, Oklahoma Medical Research Foundation (JAM)
1980–1983	Advisory Board, Environmental Cancer Information Center, City University of New York (JAM)
1981	Fellow, American Academy of Arts and Sciences (ECM & JAM)
1981	Allan D. Bass Lectureship, Department of Pharmacology, Vanderbilt University (ECM & JAM)
1981–1984	Member, Council of the National Academy of Sciences (ECM)
1982	Doctor of Science (*honoris causa*), Medical College of Wisconsin (ECM and JAM individually and separately)
1982	Princess Takamatsu Cancer Research Fund Lecturer, Tokyo (ECM & JAM)
1983	Walter Huber Lecture, British Association for Cancer Research, York, England (JAM)
1984–1986	Member, Mott Committee and Awards Assembly, General Motors Cancer Research Foundation (ECM)

1984–1987	Member, Advisory Committee to Director, National Institutes of Health (ECM)
1983	Fellow, Wisconsin Academy of Sciences, Arts and Letters (ECM and JAM individually)
1985	Honorees, Third International Symposium on Biological Reactive Intermediates, University of Maryland (ECM & JAM)
1986	Research Recognition Award, Samuel Roberts Noble Foundation, Ardmore, Oklahoma (ECM & JAM)
1992	Arkansas Toxicology Symposium Honoring the Research Contributions of Elizabeth C. Miller, Ph.D. and James A. Miller, Ph.D.

APPENDIX 2

Scientific Publications of Elizabeth Cavert Miller and James Alexander Miller

There were 270 publications between 1945 and 1987 written by Elizabeth Miller alone, with James Alexander Miller, and/or with others. This is an unwieldy list which would take a minimum of 13 pages printed in small type. Few readers would consult the list and probably even fewer would travel to a biomedical library to read the impressive collection. In lieu of the complete bibliography, what follows below are seven references which are thought to contain, in the aggregate, most or all of Elizabeth's and James's scientific papers (see references cited at end of each paper).

Miller, J.A., E.C. Miller. 1953. The carcinogenic aminoazo dyes. *Advances in Cancer Research* 1:339–396.

Miller, E.C., J.A. Miller. 1955. Biochemical investigations on hepatic carcinogenesis. *Journal of the National Cancer Institute* 15:1571–1590.

———. 1976. The metabolism of chemical carcinogens to reactive electrophiles and their possible mechanisms of action in carcinogenesis. In: *Chemical Carcinogens*, CE Searle (ed). American Chemical Society, Washington, D.C. pp. 737–762.

———. 1979. Overview on the relevance of naturally occurring carcinogens, promoters, and modulators of carcinogenesis in human cancer. In: *Naturally Occurring Carcinogen-,Mutagens and Modulators of Carcinogenesis*, E.C. Miller *et al.* (eds), Japan Scientific Societies Press, Tokyo, pp. 1–17.

———. 1979. Perspectives in the Metabolism of Chemical Carcinogens. In: *Environmental Carcinogenesis: Occurrence, Risk Evaluation and Mechanisms.* P. Emmelot, E. Kriek (eds), Elsevier/North-Holland Biomedical Press, Amsterdam, pp. 25–50.

Miller, J.A., Swanson, A.B., Miller, E.C. 1979. The Metabolic Activation of Safrole and Related Naturally Occurring Alkenylbenzenes in Relation to Carcinogenesis by These Agents. In: *Naturally Occurring Carcinogens-Mutagens and Modulators of Carcinogenesis.* E.C. Miller *et al.* (eds), Japan Scientific Societies Press, Tokyo, pp. 111–125.

Miller, J.A. 1993. Elizabeth Cavert Miller (1920–1987). In: *Women in Chemistry and Physics: A Biobibliographic Sourcebook.* L.S. Grinstein, R.K. Rose, M.H. Rafailovich (eds), Greenwood Press, Westport, Connecticut.

 | DOI 10.1163/9789004286801_011

APPENDIX 3

A Gallery of *Cancer Research* Covers

This book is a record of the accomplishments in chemical carcinogenesis of Elizabeth Cavert Miller and James Alexander Miller. The following pages with the images of five covers of *Cancer Research*, a publication of the American Association for Cancer Research, Inc., are contemporary recognition of those extraordinary accomplishments. One indication of the significance of the covers of *Cancer Research* is that the cover editor, Michael B. Shimkin (with the help of knowledgeable colleagues), become so accomplished in recognizing out of the ordinary skills and accomplishments in cancer researchers that he was asked to write (and edit) several books that are useful to both teachers and researchers. These include Shimkin, M.B., 1977. *Contrary to Nature.* Bethesda, Maryland: U.S. Department of Health, Education and Welfare, Public Health Service, National Institutes of Health (Publication NIH 76-720) and Shimkin, M.B., 1980. *Some classics of Experimental Oncology, 50 Selections, 1775–1965*. Bethesda, Maryland: U.S. Department of Health, Education and Welfare, Public Health Service, National Institutes of Health (Publication NIH 80-2150). Clearly, Elizabeth and James were “out of the ordinary”.

 | DOI 10.1163/9789004286801_012

Cancer Research, *Vol. 39, Number 11, November 1979. Recipients of 1978 and 1979 Bristol-Myers Award for distinguished Achievement in Cancer Research. The award included a $25,000 cash prizes. Elizabeth Cavert Miller and James A Miller (upper left) were recognized for contributions to the understanding of metabolism and action of chemical carcinogens. Also on the cover were Werner and Gertrude Henle, University of Pennsylvania (lower left), Mary Lasker of New York City (upper circle), and Benno C. Schmidt, also of New York (lower circle).*

COVER IMAGE COURTESY OF THE AMERICAN ASSOCIATION FOR CANCER RESEARCH, INC.

Cancer Research

VOLUME 40 • NO. 12 CNREA 8 • PP 4335-4929

December 1980

Cancer Research, *Vol. 40, Number 12, December 1980. The General Motors Cancer Research Foundation recipients of gold medals and $100,000 to several individuals. Included were Elizabeth Cavert Miller and James A Miller (fourth and fifth on right) who received the Charles S. Mott Prize for their research on the metabolic activation and covalent binding of chemical carcinogens to large biological molecules. Others in the photograph are Isaac Berenblum, Rehovot, Israel (second from left) and Elwood V. Jensen, University of Chicago (to the left of the Millers).*

COVER IMAGE COURTESY OF THE AMERICAN ASSOCIATION FOR CANCER RESEARCH, INC.

Cancer Research, *Vol. 48, Number 4, February 15, 1988. Obituary: this Cancer Research cover honors the memory of Elizabeth Cavert Miller. She was described in the cover legend as "an exemplary human being in every way."*

COVER IMAGE COURTESY OF THE AMERICAN ASSOCIATION FOR CANCER RESEARCH, INC.

Cancer Research, *Vol. 49, Number 15, August 1, 1989. Photograph of James and Elizabeth Cavert Miller made in 1985. The cover honors James A. Miller on his election as an honorary member of the American Association for Cancer Research.*

COVER IMAGE COURTESY OF THE AMERICAN ASSOCIATION FOR CANCER RESEARCH, INC.

Cancer Research, *Vol. 50, Number 6, March 15, 1990. A cover devoted to researchers on chemical carcinogenesis. Illustrated are James A. and Elizabeth Cavert Miller (upper right) as well as their former student and distinguished colleague Allan H. Conney, Rutgers (bottom right), Herbert K. Remmer, Tübingen (bottom left) and Ryuichi Kato, Tokyo (upper left).*

COVER IMAGE COURTESY OF THE AMERICAN ASSOCIATION FOR CANCER RESEARCH, INC.

APPENDIX 4

Substances Listed in the Thirteenth Report on Carcinogens

Source of the following list of carcinogen: NTP (National Toxicology Program) 2014, *Report on Carcinogens, Thirteenth Edition*. Research Triangle Park, NC: US Department of Health and Human Service. http://ntp.niehs.nih.gov/pubhealth/roc/roc13/

Bold entries indicate new or changed listings in the Thirteenth Report on Carcinogens.

Known to be Human Carcinogens

Aflatoxins
Alcoholic Beverage Consumption
4-Aminobiphenyl
Analgesic Mixtures Containing Phenacetin (see Phenacetin and Analgesic Mixtures Containing Phenacetin)
Aristolochic Acids
Arsenic and Inorganic Arsenic Compounds
Asbestos
Azathioprine
Benzene
Benzidine (see Benzidine and Dyes Metabolized to Benzidine)
Beryllium and Beryllium Compounds
Bis(chloromethyl) Ether and Technical-Grade Chloromethyl Methyl Ether
1,3-Butadiene
1,4-Butanediol Dimethanesulfonate
Cadmium and Cadmium Compounds
Chlorambucil
1-(2-Chloroethyl)-3-(4-methylcyclohexyl)-1-nitrosourea (see Nitrosourea Chemotherapeutic Agents)
Chromium Hexavalent Compounds
Coal Tars and Coal-Tar Pitches
Coke-Oven Emissions
Cyclophosphamide
Cyclosporin A
Diethylstilbestrol

 | DOI 10.1163/9789004286801_013

Dyes Metabolized to Benzidine (Benzidine Dye Class) (see Benzidine and Dyes Metabolized to Benzidine)
Erionite
Estrogens, Steroidal
Ethylene Oxide
Formaldehyde
Hepatitis B Virus
Hepatitis C Virus
Human Papillomaviruses: Some Genital-Mucosal Types
Melphalan
Methoxsalen with Ultraviolet A Therapy
Mineral Oils: Untreated and Mildly Treated
Mustard Gas
2 Naphthylamine
Neutrons (see Ionizing Radiation)
Nickel Compounds (see Nickel Compounds and Metallic Nickel)
Radon (see Ionizing Radiation)
Silica, Crystalline (Respirable Size)
Solar Radiation (see Ultraviolet Radiation Related Exposures)
Soots
Strong Inorganic Acid Mists Containing Sulfuric Acid
Sunlamps or Sunbeds, Exposure to (see Ultraviolet Radiation Related Exposures)
Tamoxifen
2,3,7,8-Tetrachlorodibenzo-*p*-dioxin
Thiotepa
Thorium Dioxide (see Ionizing Radiation)
Tobacco Smoke, Environmental (see Tobacco-Related Exposures)
Tobacco Smoking (see Tobacco-Related Exposures)
Tobacco, Smokeless (see Tobacco-Related Exposures)
***o*-Toluidine**
Ultraviolet Radiation, Broad-Spectrum (see Ultraviolet Radiation Related Exposures)
Vinyl Chloride (see Vinyl Halides [selected])
Wood Dust
X-Radiation and Gamma Radiation (see Ionizing Radiation)

Reasonably Anticipated to be Human Carcinogens

Acetaldehyde
2-Acetylaminofluorene
Acrylamide

Acrylonitrile
Adriamycin
2-Aminoanthraquinone
o-Aminoazotoluene
1-Amino-2,4-dibromoanthraquinone
2-Amino-3,4-dimethylimidazo[4,5-*f*]quinoline (see Heterocyclic Amines [Selected])
2-Amino-3,8-dimethylimidazo[4,5-*f*]quinoxaline (see Heterocyclic Amines [Selected])
1-Amino-2-methylanthraquinone
2-Amino-3-methylimidazo[4,5-*f*]quinoline (see Heterocyclic Amines [Selected])
2-Amino-1-methyl-6-phenylimidazo[4,5-*b*]pyridine (see Heterocyclic Amines [Selected])
Amitrole
o-Anisidine and Its Hydrochloride
Azacitidine
Basic Red 9 Monohydrochloride
Benz[*a*]anthracene (see Polycyclic Aromatic Hydrocarbons: 15 Listings)
Benzo[*b*]fluoranthene (see Polycyclic Aromatic Hydrocarbons: 15 Listings)
Benzo[*j*]fluoranthene (see Polycyclic Aromatic Hydrocarbons: 15 Listings)
Benzo[*k*]fluoranthene (see Polycyclic Aromatic Hydrocarbons: 15 Listings)
Benzo[*a*]pyrene (see Polycyclic Aromatic Hydrocarbons: 15 Listings)
Benzotrichloride
2,2-Bis(bromomethyl)-1,3-propanediol (Technical Grade)
Bis(chloroethyl) Nitrosourea (see Nitrosourea Chemotherapeutic Agents)
Bromodichloromethane
1-Bromopropane
Butylated Hydroxyanisole
Captafol
Carbon Tetrachloride
Ceramic Fibers (Respirable Size)
Chloramphenicol
Chlorendic Acid
Chlorinated Paraffins (C_{12}, 60% Chlorine)
Chloroform
1-(2-Chloroethyl)-3-cyclohexyl-1-nitrosourea (see Nitrosourea Chemotherapeutic Agents)
3-Chloro-2-methylpropene
4-Chloro-*o*-phenylenediamine
Chloroprene
p-Chloro-*o*-toluidine and Its Hydrochloride
Chlorozotocin (see Nitrosourea Chemotherapeutic Agents)

Cisplatin
Cobalt Sulfate
Cobalt–Tungsten Carbide: Powders and Hard Metals
p-Cresidine
Cumene
Cupferron
Dacarbazine
Danthron
2,4-Diaminoanisole Sulfate
2,4-Diaminotoluene
Diazoaminobenzene
Dibenz[*a,h*]acridine (see Polycyclic Aromatic Hydrocarbons: 15 Listings)
Dibenz[*a,j*]acridine (see Polycyclic Aromatic Hydrocarbons: 15 Listings)
Dibenz[*a,h*]anthracene (see Polycyclic Aromatic Hydrocarbons: 15 Listings)
7H-Dibenzo[*c,g*]carbazole (see Polycyclic Aromatic Hydrocarbons: 15 Listings)
Dibenzo[*a,e*]pyrene (see Polycyclic Aromatic Hydrocarbons: 15 Listings)
Dibenzo[*a,h*]pyrene (see Polycyclic Aromatic Hydrocarbons: 15 Listings)
Dibenzo[*a,i*]pyrene (see Polycyclic Aromatic Hydrocarbons: 15 Listings)
Dibenzo[*a,l*]pyrene (see Polycyclic Aromatic Hydrocarbons: 15 Listings)
1,2-Dibromo-3-chloropropane
1,2-Dibromoethane
2,3-Dibromo-1-propanol
1,4-Dichlorobenzene
3,3′-Dichlorobenzidine and Its Dihydrochloride
Dichlorodiphenyltrichloroethane
1,2-Dichloroethane
Dichloromethane
1,3-Dichloropropene (Technical Grade)
Diepoxybutane
Diesel Exhaust Particulates
Di(2-ethylhexyl) Phthalate
Diethyl Sulfate
Diglycidyl Resorcinol Ether
3,3′-Dimethoxybenzidine (see 3,3′-Dimethoxybenzidine and Dyes Metabolized to 3,3′-Dimethoxybenzidine)
4-Dimethylaminoazobenzene
3,3′-Dimethylbenzidine (see 3,3′-Dimethylbenzidine and Dyes Metabolized to 3,3′-Dimethylbenzidine)
Dimethylcarbamoyl Chloride
1,1-Dimethylhydrazine

Dimethyl Sulfate
Dimethylvinyl Chloride
1,6-Dinitropyrene (see Nitroarenes [Selected])
1,8-Dinitropyrene (see Nitroarenes [Selected])
1,4-Dioxane
Disperse Blue 1
Dyes Metabolized to 3,3′-Dimethoxybenzidine (3,3′-Dimethoxybenzidine Dye Class) (see 3,3′-Dimethoxybenzidine and Dyes Metabolized to 3,3′-Dimethoxybenzidine)
Dyes Metabolized to 3,3′-Dimethylbenzidine (3,3′-Dimethylbenzidine Dye Class) (see 3,3′-Dimethylbenzidine and Dyes Metabolized to 3,3′-Dimethylbenzidine)
Epichlorohydrin
Ethylene Thiourea
Ethyl Methanesulfonate
Furan
Glass Wool Fibers (Inhalable), Certain
Glycidol
Hexachlorobenzene
Hexachloroethane
Hexamethylphosphoramide
Hydrazine and Hydrazine Sulfate
Hydrazobenzene
Indeno[1,2,3-*cd*]pyrene (see Polycyclic Aromatic Hydrocarbons: 15 Listings)
Iron Dextran Complex
Isoprene
Kepone
Lead and Lead Compounds
Lindane, Hexachlorocyclohexane (Technical Grade), and Other Hexachlorocyclohexane Isomers
2-Methylaziridine
5-Methylchrysene (see Polycyclic Aromatic Hydrocarbons: 15 Listings)
4,4′-Methylenebis(2-chloroaniline)
4,4′-Methylenebis(*N,N*-dimethyl)benzenamine
4,4′-Methylenedianiline and Its Dihydrochloride
Methyleugenol
Methyl Methanesulfonate
N-Methyl-*N′*-Nitro-*N*-Nitrosoguanidine (see *N*-Nitrosamines: 15 Listings)
Metronidazole
Michler's Ketone
Mirex
Naphthalene

Nickel, Metallic (see Nickel Compounds and Metallic Nickel)
Nitrilotriacetic Acid
o-Nitroanisole
Nitrobenzene
6-Nitrochrysene (see Nitroarenes [Selected])
Nitrofen
Nitrogen Mustard Hydrochloride
Nitromethane
2-Nitropropane
1-Nitropyrene (see Nitroarenes [Selected])
4-Nitropyrene (see Nitroarenes [Selected])
N-Nitrosodi-*n*-butylamine (see *N*-Nitrosamines: 15 Listings)
N-Nitrosodiethanolamine (see *N*-Nitrosamines: 15 Listings)
N-Nitrosodiethylamine (see *N*-Nitrosamines: 15 Listings)
N-Nitrosodimethylamine (see *N*-Nitrosamines: 15 Listings)
N-Nitrosodi-*n*-propylamine (see *N*-Nitrosamines: 15 Listings)
N-Nitroso-*N*-ethylurea (see *N*-Nitrosamines: 15 Listings)
4-(*N*-Nitrosomethylamino)-1-(3-pyridyl)-1-butanone (see *N*-Nitrosamines: 15 Listings)
N-Nitroso-*N*-methylurea (see *N*-Nitrosamines: 15 Listings)
N-Nitrosomethylvinylamine (see *N*-Nitrosamines: 15 Listings)
N-Nitrosomorpholine (see *N*-Nitrosamines: 15 Listings)
N-Nitrosonornicotine (see *N*-Nitrosamines: 15 Listings)
N-Nitrosopiperidine (see *N*-Nitrosamines: 15 Listings)
N-Nitrosopyrrolidine (see *N*-Nitrosamines: 15 Listings)
N-Nitrososarcosine (see *N*-Nitrosamines: 15 Listings)
o-Nitrotoluene
Norethisterone
Ochratoxin A
4,4′-Oxydianiline
Oxymetholone
Pentachlorophenol and By-products of Its Synthesis
Phenacetin (see Phenacetin and Analgesic Mixtures Containing Phenacetin)
Phenazopyridine Hydrochloride
Phenolphthalein
Phenoxybenzamine Hydrochloride
Phenytoin and Phenytoin Sodium
Polybrominated Biphenyls
Polychlorinated Biphenyls
Procarbazine and Its Hydrochloride
Progesterone

1,3-Propane Sultone
β-Propiolactone
Propylene Oxide
Propylthiouracil
Reserpine
Riddelliine
Safrole
Selenium Sulfide
Streptozotocin (see Nitrosourea Chemotherapeutic Agents)
Styrene
Styrene-7,8-oxide
Sulfallate
Tetrachloroethylene
Tetrafluoroethylene
Tetranitromethane
Thioacetamide
4,4′-Thiodianiline
Thiourea
Toluene Diisocyanates
Toxaphene
Trichloroethylene
2,4,6-Trichlorophenol
1,2,3-Trichloropropane
Tris(2,3-dibromopropyl) Phosphate
Ultraviolet Radiation A (see Ultraviolet Radiation Related Exposures)
Ultraviolet Radiation B (see Ultraviolet Radiation Related Exposures)
Ultraviolet Radiation C (see Ultraviolet Radiation Related Exposures)
Urethane
Vinyl Bromide (see Vinyl Halides [Selected])
4-Vinyl-1-cyclohexene Diepoxide
Vinyl Fluoride (see Vinyl Halides [Selected])

Index